Financial Markets and Martinga

Springer
London
Berlin
Heidelberg
New York
Hong Kong
Milan
Paris
Tokyo

Nicolas Bouleau

Financial Markets and Martingales

Observations on Science and Speculation

Translated by Alan Thomas

With 14 Illustrations

 Springer

Nicolas Bouleau, PhD, Alumnus of the École Polytechnique
École des Ponts, 28 rue des Saint Pères, 75007 Paris, France

Translator
Alan D. Thomas
Department of Mathematics, University of Wales Swansea, Singleton Park, Swansea,
SA2 8PP, Wales, UK

Ouvrage publié avec le concours du Ministère français chargé de la culture – Centre
national du livre.

Published with the assistance of the French Ministry of Culture – CNL.

British Library Cataloguing in Publication Data
Bouleau, Nicolas
    Financial markets and martingales.
    1. Capital market – Mathematical models 2. Martingales (Mathematics)
    I. Title
    332.6'01519287
ISBN 1852335823

Library of Congress Cataloging-in-Publication Data
Bouleau, Nicolas.
    [Martingales et marchés financiers. English]
    Financial markets and martingales / Nicolas Bouleau ; translated by Alan Thomas.
        p. cm.
    Includes bibliographical references and index.
    ISBN 1-85233-582-3 (alk. paper)
    1. Portfolio management. 2. Investment analysis. 3. Martingales (Mathematics) I. Title.
HG4529.5.B6913 2004
332.6'01'519236—dc22   2003061741

ISBN 1-85233-582-3 Springer-Verlag London Berlin Heidelberg
Springer-Verlag is a part of Springer Science+Business Media
springeronline.com

Typesetting: Camera-ready by Alan Thomas and Thomas Unger
12/3830-543210 Printed on acid-free paper SPIN 10856966

Contents

Part IV. The Stakes and the Payoffs

Introduction

This book is not a text book for the operators in market rooms, nor a work of ideological propaganda. It is simply the observations of a mathematics professor from an engineering college, who has worked with banks on new financial products. I have taken part in a particular experiment, where certain questions have frequently been asked.

There is one, with which I intend to start. Is it really possible to make money in the financial markets?

Some people say the best you can do is to win in a portion of the market, without taking increased risks.

This is correct perhaps, but it is not the full story. Give up any hope of certainty. Let us not pursue this obvious reply too closely because of this striking fact: in economics, the prediction of values from the various *theories* is even weaker than in the physical sciences.

I think it is instructive to tackle this question in a *concrete* way. In other words, the best way to penetrate the mysteries of finances seems to me to be to develop an opinion progressively by taking the journey oneself. You begin with a blank piece of paper, list several portfolios, and methodically experiment with each of them, day by day, with the ideas that are essential to their management. For this you will need to keep yourself informed about the prices, to record them in your journal, to write commentaries on the values concerned, and to append to them any economic facts and opinions, since opinion does have its importance. You can considerably enlarge your field of strategies if you can test them with a computer and a spreadsheet. These allow you to calculate statistics, annual or monthly, moving averages, variance, and to study the correlation between various data, and confront any economic significance. The graphical possibilities allow you to test the geometric flavour of your deductions.

By experimenting over several months, by discovering many things that are not in the financial press, a semblance follows. You learn from your success and failure. You can go further. After the management of portfolios of transferable securities, you can undertake those of raw materials or currencies and move finally onto derivatives, futures and options. These last are the true instruments of taking a position: because with them you can construct different bets on the future. To do this exercise you do not need to pay a subscription for instantaneous quotes, even if you want to test the most recent methods of hedging. That is to say, protection against risk[1].

This practical approach has the advantage of treating finance as an open book, which it really is. Finance more or less reflects social life through the medium of economic facts and by the interpretation made by the players on the stock markets. They express their views by calculations and with models, and it is interesting to know that for about twenty-five years the daily operations of the markets relies, twenty-four hours a day, on recent mathematical research.

New practices appear, and the old methods of analysing the activities of finance turn out to be unsuitable today. Precisely because it is difficult to schematise its operation, the importance taken by finance in various businesses and in the economy world-wide merits being carefully examined.

It is clearly a highly polemical subject. It questions justice and morality from the aspect of financial questions, and the reader's point of view changes. Good sense tells us to stop being naive, and to see the world as a polarisation of forces and interests. Money is a substitute which in itself has no a real significance other than a means to buy goods. Our interpretation of its daily movements reflects on the mythical sources of our culture. That is how we have freedom of price, for example. Why is the price of bread free? Starting from bread, the people of Paris asked for Madame Veto[2]. The rights of man against free trade? We see the historical and political process in full. We must be careful not to get stuck in the mud.

Finance is above all practical. But, like every technical subject, it is influenced by elaborate ideas for self-improvement. Until the Second World War, it was considered a discipline quite distinct from economics. It was taught in an essentially descriptive way, putting accent on the institutional and legal aspects, and on day to day calculations. During the third quarter of the 20th century, it became the object of a structured and argued economic theory with all the variations and controversies that are known in science. This development was principally the fruit of the American university schools of economics, with a

---

[1] The Internet sites of the organised markets (see "Organised markets" in the Glossary, page 140) give a large amount of free information about quotations.

[2] Louis XVIth and Marie Antoinette were nicknamed Monsieur and Madame Veto by the Republicans because the Constituent Assembly allowed the king to have the power to veto any decree submitted to him. (Translator)

significant contribution from the French school[3]. This then has been developed: the theory of the efficient market, the theory of portfolio selection, the analysis of risk, models of the market in the manner of the Markowitz-Tobin model, which makes use of time in a most simple way, over a period, and of risk by expectation and variance and then the *CAPM (Capital Asset Pricing Model)* in a version of equilibrium, and their various improvements.

The practitioners of finance today cannot ignore this large conceptual body, which uses all sorts of mathematics (probability and statistics, control theory, optimisation, etc.). This use of mathematics by financial economics raises a delicate question because of the problems of rigour and of generality, that lead the researchers to drive for very deductive developments. But can a very deductive theory be legitimised by social science? Mathematicians inevitably seek a new perspective on those hypotheses that provide a possible line of argument. The economists then try to chose the best symbolic representations to pursue these questions. It is normal to accept more mathematics in areas where there is a close connection between applications and the use by practitioners. This is the case in finance.

Today this is what the traders do in the finance markets in Chicago, Tokyo, London, Paris or Singapore, with consequences in commerce and for the global economy which is thought of in terms of *Itô calculus* and of several other ideas from this area of mathematics called *stochastic calculus*[4]. From the point of view of the scientific discipline (financial economy) it is a relatively minor point that an intimate connection is found between very advanced mathematics and the work of the traders on the market floors of financial institutions. But this point is crucial. It is about the question of evaluation and the hedging of *options*[5]; that is to say, the problem of calculating their price with good control against risks. On this precise point, a very strong logic, for which the work of economists in universities has supplied the arguments, has made the tools of mathematical theories and abstractions both concrete and operational. This connection between mathematics and the financial markets allows mathematicians and bankers to work together.

New practices appear. We are going to study these, and attempt to clarify certain situations. It seems to me that a pertinent investigation should tackle various questions about the role of financial markets without which it would be necessary to refer first to the theories of economic finance. These provide a

---

[3] Amongst the pioneers, we must mention Arrow, Debreu, Allais, Lintner, Markowitz, Miller, Modigliani, Samuelson, Sharpe, Tobin, etc.

[4] Part of the theory of probability that concerns random processes and the associated calculations.

[5] See "Options" in the Glossary, page 138.

useful insight, and there are good books in French[6] and also in English[7], which
I am pleased to recommend to the reader. As for the modelling of stability of
the choice of agents in the presence of risk, which we call the *theory of portfolio
selection*, I think that this has taken on an excessive importance considering
the brittleness of the hypotheses (rational behaviour of agents and the utility
function, economic thinking of the market), and above all the existence of an
equally general probability governing the future point of view for everyone, etc.
I will present some thoughts on the markets and on the behaviour of foreign
exchange dealers without using these hypotheses and these theories.

*

* *

These new ideas that allowed the development of the derivatives markets[8] and
which are the application of advanced mathematics, appeared in the United
States in the 1970s. But in France, the coming together of bankers and mathematicians did not happen until the mid-1980s. This story is worth telling.

In order to take account of the disruption and distress caused by this, it
is necessary to realise that the intellectual life of a Parisian was dominated by
the ideas of the left. That is why the community of French mathematicians
had as common denominator a certain idealism of the discipline, and a mis-
trust of applications. Mathematics must not be put into the service of the
military-industrial complex, and must be denied to the nuclear industry and
to capitalism. Bourbaki set the tone with his monumental work in pure mathematics.

Nevertheless the applications that one uses here and there, more and more,
were considered as a source of work destined to complement the resources of
research teams. In truth, the applications were often used thanklessly. Rarely
were there contracts with industry that allowed the use of advanced or recent
mathematical works. It is over eighty years since the banks started to be seri-
ously interested in recruiting young people with a scientific background, notably
from among the graduates of engineering faculties, not to convert them towards
management, but to use their mathematical expertise.

In 1985, a young engineer recruited by a large merchant bank attracted
the attention of my research team because of reviews of his articles for financiers which were well received. The concepts of stochastic calculus were
freshly elaborated in Pure Mathematics. What can interest the bankers in the

---

[6] In order of increasing technicality we cite: Guesnerie [47], Aglietta [3], Dumas and
Allaz [36], and for those who enjoy anecdotes, Bernstein [11].
[7] Huang and Litzenberger [50], Copeland and Weston [26] Ingersoll [52].
[8] As one calls the financial markets of derivatives (see "MATIF" and "Derivative"
in the Glossary, pages 138 and 135).

Itô integral on which we have worked for many years? Was it serious or was it just a vanity of scientists, as one sometimes finds?

We set to work theoretically at first, and then we will consider the influence of the *trader*[9] and of the working relationships that are woven into the market rooms. The MATIF opened in 1986. Some colleagues in the Université Paris-Dauphine had already been working on these problems, that were worrying other establishments but which had been under vigorous investigation in the United States for more than a dozen years.

Now I embark on an unexpected intellectual adventure that was not only serious in terms of research, but was also a turning point in the history of applied mathematics. It was to create a strong movement as much for research as for professional education. Clearly in the start-up years of financial mathematics the ethical questions were largely discussed in the heart of the research community. To place our activities at the heart of the temples of profit poses problems of conscience. Nevertheless, we will be sensitive to the fact that it is not about one application out of all the others – modelling then numerical calculations – but of the discovery of a profound link, a new comprehension, that encourages us to follow, and to see.

This theme is now treated in higher education in a number of countries, fed by a regular flow of students. Ironically, a few years ago, it was considered perfectly normal for the Ecole Normale, which had been in Paris one of the high altars of Marxist thought, to organise one of the most prestigious international colloquia on derivatives.

*
* *

The reader of this book needs no previous knowledge of finance, economics or mathematics. No complicated equations that serve the market traders figure in this work. I have tried to put the principal ideas into perspective, to articulate them and to list the pending questions. I wish to allow the reader, who will have some knowledge of the new relations between mathematics and finance, to move easily into this complex area and to refine his vision of things.

A glossary can be found at the end of this book. This is for information rather than for explanation and its postponement is not necessary for comprehension of the text. If there are obscurities here and there are, they are due to errors or incomprehension in my part, for which I ask, in advance, the indulgence of the reader.

We start our journey in the gaming rooms of the casinos. It was from the games of chance that probability theory was born, and from which the

---

[9] The name given to operators in the market rooms of banks who buy and sell products, shares, assets, etc.

mathematical tools used in finance have been forged. Moreover, this practical introduction corresponds well to the *human* side of finance. The serious tone of journals and practitioners must not mislead you, for the gamblers in the casinos are themselves also serious. This happens as if finance was a real passionate game, mobilising all the mental faculties, open to extremely varied initiatives, and in which, a sufficiently rare situation, intelligence is very clearly the source of power. The games of chance are scaled down versions of this. Mathematics can rend them perfectly to account.

One can use the mathematical language of chance to talk about prices on the stock market as Louis Bachelier did at the start of the 20th century. This connection is not obvious, except when one recalls the complexity of economic phenomena and the concept of chance that Henri Poincaré had developed for dynamical systems. The work of Bachelier, ignored for a long time, is considered today as pioneering modern finance. It allows us to talk about Brownian motion, a mathematical object central to many areas, as well as the random processes that follow. These one does not handle with the usual rules of differential calculus (as elaborated by Leibniz and Newton) but with *Itô calculus*, which has become routine in the market floor. This unfortunately accentuates the esoteric character of finance.

Once we notice that the gains from a sequence of buying and selling certain amounts of assets can be written as an *integral*, that is to say as a general sum or a continuous sum, we will know enough to tackle derivatives, whose management uses these new mathematical techniques.

We arrive here at the heart of the problems of finance, created by the conceptual revolution of the 1970s. It was by this method of the treatment of risks and futures contracts that the existence of organised markets shattered the economic way of thinking.

According to the methods in use in the 1960s, when a bank sold a futures contract of say 3 months, it proposed a price of the *expected value* for the security, augmented by a beneficial margin, namely the average of the probable value over the 3 months. It then called on its experts in economic analysis to evaluate these probabilities. Once the contract of sale had been made, nobody worried until 3 months later, when they would see whether it had lost or gained, and by how much.

The epistemological rupture of the 1970s consisted of noticing that the buying and selling of securities in suitable quantities during these 3 months would allow the bank to realise the contract exactly, at a set price that one could calculate at the beginning. This fixed amount must necessarily be the price of the contract, because if the transaction were made at a different price, whether better for the bank or the client, somebody would realise a profit. In other words, it is the only price that prevents all arbitrage. It prevents all possibility of profit without risk. The market exposes itself to the risk it fears so much by taking account of the market at each instant, so that one can really *abolish risk*. It is a mathematical miracle. This allows the exact hedging of a contract. By proceeding in this way a bank exploits the possibilities offered by the organised markets that supply it continuously with an instantaneous quotation to buy or sell. The principle of hedging of contracts has an undeniable logic. It allowed the rapid development of derivatives and their corresponding organised markets.

It is about a new way of thinking about the management of risks and financial anticipation. It requires the constitution of a hedging portfolio according to a fairly sophisticated technique. It gives a greater importance to information supplied by the markets, but a lesser importance to the expertise of forecasters. This is the change that seems to me the most considerable: *the listening to the market becomes the main objective reality.*

Does the knowledge of the operation of financial markets and of the mathematical paraphernalia that they use allow us to throw away the general dossier on financial speculation? It is about natural heterogeneous practices without frontiers. An image appears good enough: one can summarise the recent development by saying that finance has become *high-tech*. As in the high-tech industries such as Information Technology and Genetic Engineering, the science of selling is a condition of profits. In finance as in these sectors, the science is enriched by this practice, all of which necessarily leads to ethical problems that one must explain.

In the last Part, I take on board the theme of the recent increase in financial powers and of the legitimacy of this development. The setting-up of derivatives markets plays a determining role here. This role is, with respect to the economy, similar to the role of the media with respect to opinion. The exercise of financial power is characterised by the absence of inertia and by violent fractures like those in fragile materials. In comparing the procedures of decision making, bringing in the effect of political force, in the areas of land management or the environment it seems to me that the financial powers act in a sufficiently wild fashion so that relatively futile causes can have effects on a great scale. One can hope that the construction of Europe will be a privileged occasion for experimentation and improvement.

Part I

Martingales

1
Luck in the Casino

He always wins, I noticed it only yesterday. He must have a winning formula, and I always play the same as he does. Yesterday he won again as always, but I made the mistake of continuing to play after he had left: that was my fault.

Stefan Zweig, Twenty-four Hours in the Life of a Woman[1].

The Passion for Gambling

Excuse me my angel, but I would like to tell you about my business, so you can clearly see the stakes at play. Here you are: more than twenty times I have tested this by playing. If you proceed cold-bloodedly, calmly and methodically, there is no way you can lose. I swear to you, no way! There, blind chance, that's my method. [66]

This is what Dostoïevski wrote to Anna Grigorievna Snitkine. He had just married her after the death of his first wife, and they were on their honeymoon. He left her at Dresden and went alone to Hamburg to play the casino.

Several months before, in *The Gambler*, the author had perceptively analysed the fascination of the game. Not only did he retain his illusions during the course of this work, but he also taught them to his wife in the novel itself.

---

[1] ©1980 Williams Verlag AG, Zurich/Atrium Press Ltd., London.

She knew all the details because she had taken down his notes in shorthand, and it was in doing this that love blossomed between them. She was forgiving and anxious at the same time. Several letters describe in detail how Fedia had already lost the money he had taken with him, the money she had sent him, and that he had forfeited his watch.

Consequently, I have this trump in reserve. But how did it happen? [66]

Dostoïevski explains that his passion had overpowered him, and did not know how to defeat it. He had not been able to apply his method.

I was cold-blooded and reserved, my nerves were tingling, I started to gamble and started to lose my temper and to put money down without calculating, and then losing. Any man who plays without calculating, taking chances, is a fool. [66]

This paragraph from one of our most profound literary minds is startling. When he wrote that you must not play by chance the games of chance, he stated one of the key problems affecting economics and modern finance. But, by saying to his loved-one that he played for high stakes and was fired with enthusiasm because he loved her, and that he wanted her back as soon as possible, he did no more than push her into the trap that he had made for himself.

I have made super-human efforts to remain calm and methodical for a whole hour, and the result was that I won thirty frederics of gold, that is to say three hundred florins. [66]

These would soon be lost, along with much more. Anna had to endure several false announcements of his return, and make humiliating trips to the bankers, to other towns as well, to send him more money he said he needed for the journey.

Fiodor Mikhaïlovitch stopped gambling completely after forty years, as if either the pleasure of gambling had completely vanished, or he had become disenchanted. In 1871 he wrote:

I was enslaved by the game; from now on I live only for my work. My nights will no longer be spent dreaming of infallible martingales. [66]

He kept his word.

The Martingales of Gamblers

What are these *martingales* that enslave gamblers? The word comes from *martingale chausses*, a medieval form of culottes that can be opened from behind, similar to garments worn by the inhabitants of Martigues in Provence. M. Littré tells us that King François I in his bravery

> ... *was of such constitution, that every time he wanted to go to battle, it was necessary that he arranged his clothes and dismounted his horse to answer a call to nature; and for this he ordinarily wore martingale chausses*[2].

The term was adopted within gambling as a metaphor for the intersection of the harness belt that was called a martingale. Probably the success of this expression comes from the success of the design. Since the 18th century it has come to mean a fitted fold in the back of a coat that cannot be seen from the front. This suggests the idea of a secret weapon, known only to the gambler and which is the reason for his perseverance.

The oldest martingale known is *doubling the stakes*. Take the game of fair roulette, that is has no tax nor zero of any sort. This is what we would call a *pure casino*[3]. By betting on something that has two equally likely outcomes, high or low, odd or even, black or red, there is an equal chance of winning or losing the amount of your bet. If when you lose you double the stake, it is easy to see that when you do win, the amount won will compensate for all your losses from the beginning, and moreover, bring a gain equal to that of your initial bet. It makes good sense that after a period of time you must win, and probability theory can confirm this, so this method makes winning a certainty.

You could instead wait until you have lost ten times in succession on an *even* before playing an *odd*. Or, if you think that you have periods of luck from which you can benefit, double the stake when you win, and divide it by two when you lose. Another way to play is to attempt to hold out for a long time by playing small stakes each time. We all know that gamblers' luck eventually comes around, so you must have patience to wait until you have regained your losses, plus the initial stake and stop there. This method, like doubling the stake, undoubtedly wins.

All martingales have a flaw. Sometimes it is in a mistaken assumption, such as the idea that in one way or other the gamblers can influence what happens,

[2] Dictionnaire de la Langue Française, E. Littré, Librairie Hachette, 1883.
[3] Such a game is characterised by the fact that, at each turn, the mathematical expectation of a win is zero. The mathematical expectation, introduced by Pascal and Fermat is the algebraic sum of the gains and losses, taking into account their probabilities.

even though a roulette ball has no memory. Sometimes they conceal practical difficulties behind clear principles, such as doubling the stake, which, as soon as a run of bad luck lasts too long, mercilessly ruins even the most confident gambler.

The wisdom that probability theory can give us towards a game is difficult to teach since martingales are numerous and diverse. I made my teaching debut to student engineers of the Ecole des Ponts. I was convinced that in order for them to take on their responsibilities when faced with natural risks, such as floods, avalanches and so on, they should be made immune to gambling by means of a straightforward practical demonstration of risk. I would give each of them a number, then select a number randomly and give a test to the one chosen. The injustice of the process was instructive, since at the end of the semester some students had never been tested, whilst others had been tested several times. On the strength of this experience I organised a computer-based practical where each student was given an artificial casino, an initial sum of money and a portfolio of martingales whose respective weaknesses they had to identify. This turned out to be an educational catastrophe! Whenever the number of students was fifty or more, there never failed to be one amongst them who won in such an outrageous way that they all concluded he had an exceptional gift. Things became farcical and it was difficult bringing reason back on track!

It may seem less interesting, but it is more convincing to undertake a detailed mathematical study of the main known or conceivable martingales. This is what probabilists Dubins and Savage[4] did, even if the more subtle martingales escape analysis. What lessons can we learn from analysing these martingales? Essentially, that there is no guaranteed method of winning. This response is nevertheless incomplete, and economics, like life in general, places us in a risky position to gamble. Since we may have to gamble without being a gambler it is interesting to know, having admitted that there is no winning strategy, if there are dangerous or inept ways of gambling, and what differentiates between different strategies.

To approach this question we must revisit mathematics, and discuss the salient ideas on gambling and games. They appeared on the stock market in their present form in relation to economics following the pioneering work of Louis Bachelier, who introduced us to Brownian motion and to those random processes that, by an interesting development of the word martingale, we now call *mathematical martingales.*

---

[4] Dubins and Savage [32].

2
The Stock Market and Probability

Alice came to the conclusion that it was a very difficult game indeed. The players all played at once without waiting for turns...

Lewis Carroll, Alice Through The Looking Glass

Economy and Chance

The ball which the croupier throws onto the roulette table moves in a well-determined direction, with each bounce governed by the laws of mechanics. Their succession is so complex however, that the slot where it finally comes to rest is governed by chance. The laws for the calculation of probabilities govern the gamblers' winnings based on their bets.

Players in finance are numerous, and they play simultaneously. One buys here to sell here, another to export, one borrows to equip himself, another invests his own funds and so on, to the extent that the course of assets, currency and indexes are permanently in a state of flux. We are used to seeing in the financial pages of the daily papers these jagged diagrams looking like the silhouette of a mountain range. Is this turbulence random? A priori no. Each economic player is a decision-maker, with precise requirements, following well-defined objectives. But the number of players, the diversity of their interests and the complexity of their interactions are such that it is reasonable to consider that their effects on the market listings bring into play the laws of chance.

Historically, the way to calculate probability was known before the concept
of chance had been examined and clarified. From Pascal and Fermat in the 17th
Century, to Laplace and Poisson in the 19th Century, encompassing Huygens
(*De Rationibus in Ludo Alea*, 1657), Jaques Bernoulli (*Ars Conjectandi*, 1713)
and Abraham de Moivre (*The Doctrine of Chances*, 1718), increasingly detailed
reasoning and calculation was pursued in the analysis of games. This included
counting the numbers of possible and favourable cases, to try variations of
an uncompleted game (decision problems), evaluate the asymptotic tendencies
during a large number of experiments, in order to found a science of the *reason
to believe*, according to the expression used by Condorcet[1], as well as the start of
a theory of statistical inference with Gauss and Laplace[2]. However the genesis
of random phenomenon was never really questioned.

The understanding of the true nature of chance, like the result of effects
produced by a cumulative set of numerous causes, can be attributed to Henri
Poincaré. By the analysis of examples (the fall of an object in equilibrium,
meteorology, distribution of small planets on the zodiac, roulette, kinetic theory of gases, drops of rain), and by his clear mathematical explanation of card
shuffling and the distribution of decimals in logarithmic tables, he showed that
chance is the result of the development of a system sensitive to small pertur-
bations.

He expands on this subject in his book [74], opening the way to one of the
most important themes in contemporary science: the understanding of Chaos
Theory[3].

So even if each economic player makes his decisions in a deterministic manner, it is not unreasonable to consider that the development of the prices, that
represents the fluctuations of supply and demand between numerous agents
and is subject to varying constraints, is a phenomenon where chance intervenes and which presents a certain analogy with what is seen at the casino.
The Frenchman Louis Bachelier was the first, it appears, to propose the modelling of the stock market prices and provide estimations on forward purchases
using probabilistic reasoning based on gambling[4].

---

[1] See R. Rashed, *Condorcet, Mathématique et société*, Hermann, 1974, and Condorcet, Nicolas Caritat de, Article on Probability (1775) in the *Encyclopédie
Méthodique*, Vol 2. pp. 640–63 of the section Mathematics.

[2] "Théorie analytique des probabilités", 1812 and "Essai philosophique sur les probabilités", 1814.

[3] See Ekeland [38], Ruelle [77], Peters [73].

[4] I do not think that the modelling of the price by a random process is the only one
that merits being studied. I accord it a privileged place because it is rich in teaching
and in the possibilities of expressing reality. But, for example, the representation
where the dynamics of deterministic chaos and the pure risks are presented is also
topic for research.

The Ideas of Louis Bachelier

In his formal thesis in 1900, he proposed a theory which today is called *evaluation of options*, meaning the forward purchases or sales under the condition that the price is greater than or equal to a fixed value[5].

Such calculations were already known on the stock market, as well as for the other conditional transactions that Bachelier also studied. The principal hypothesis on which his argument is based is that the price of an asset set by the market is mathematically presented in the same way as the gains (the algebraic accrued winnings) of a gambler in a game of chance. By infinitesimal reasoning, that comes back to the statement that the increments of prices during successive instants are independent, he came up with a partial differential equation that he called the *equation of the radiation of probability*, formally identical to that which Joseph Fourier obtained almost a century before to describe the diffusion of heat in a homogenous body. According to these calculations, the probability that governs the future price is represented by an increasingly flattened Gaussian bell curve, the standard deviation of which increases like the square root of the time from now. These formulae, that Bachelier was to investigate in more detail in his later work, are those of mathematical Brownian motion, for which he consequently found himself to be one of the pioneers.

Henri Poincaré was a member of the board of examiners for his thesis. His unenthusiastic report shows how divided he was. On one hand, the mathematical expression by Louis Bachelier was imperfect and the logic often misleading. This was also the opinion of Kolmogorov when in 1931 he contributed to bringing Bachelier to the fore by referring to him as the creator of the theory of *instant memory* over continuous-time random processes, or Markov processes, and made the connection with the Fokker-Planck equation from physics[6].

We know today that Bachelier's arguments can lead to different results, so the criticism of his lack of precision is not purely academic. Bachelier failed to see that one could leave the theoretical hypotheses from the theory of Gaussian errors. On the other hand, Poincaré had closely examined with detailed analysis the use of normal distribution (or Gauss Law) and the least-squares method that justified it.

He writes in his book:

The world believes it because the mathematicians think that it is a fact of observation, and the observers think that it is a theory of mathematics. [74]

---

[5] We will discuss later, in Part II, the concept of an option, that we will make explicit.

[6] A.N. Kolmogorov, "Über die analytischen Methoden in der Wahrscheinlichkeitsrechnung", Mathematische Annalen, 1931.

Poincaré was not in a position at this time to completely describe which were the non-Gaussian laws that the addition of small independent random variables could generate – this result would only be established later in the 1930s by Lévy and Khinchin – but he knew some examples and had the evidence that the Gaussian hypotheses may not be satisfied[7].

However the creation of a *law of radiation of probability* by reasoning using games with infinitesimal stakes could not fail to please Henri Poincaré, since it was so much in the spirit of the analysis of the concept of chance that he produced his works on dynamical systems[8]. A summary treatment of a delicate subject is often, however, what scientists forgive the least.

Processes having the Centre of Gravity Property

To model the stock exchange, Bachelier considered that

> The market does not take account, at any given instant, of either the rise or the fall in real price[9].

Consequently, prediction presents itself as a game of chance. Without bringing any justification other than good intuitive plausibility, he added that *the expectation of the speculator is zero.* But this amounts to saying much more, since it affirms that not only is there uncertainty about the rise or fall, but that they can be accurately quantified. If you weight the prices an instant later by their probability, the *centre of gravity* of the distribution obtained is the current value.

In doing this, he introduced into mathematics a remarkable theory. (He was only interested in the specific cases of the Gaussian hypotheses providing classic bell curves). The property – *the current value is the centre of gravity of the values at a later instant weighted according to their probability* – has the virtue of being *transitive.* Any engineer knows that to calculate the centre of gravity of a body one can group together the mass of a part at its centre of gravity. The centre of gravity of the centres of gravity is still the centre of gravity.

---

[7] He cites explicitly the case of error following Cauchy's law in [74].

[8] This is related equally to his works on the theory of potential, notably in his "Théorie analytique de la chaleur", published in 1895.

[9] Théorie de la spéculation, Annales Scientifiques de l'Ecole Normale Supérieure, 17 (1900), 21–86. Reprinted in P.H. Cootner (Ed), The Random Character of Stock Market Prices, MIT Press, 1964.

The random processes that have this property are called *martingales* in modern mathematics; a term that takes on a meaning quite different from that which it has in the language of gamblers. However this theory, based on the very simple property of the centre of gravity, was only brought out much later in the 1940s. As often happens, mathematicians first studied a particular case, that of the Gaussian hypothesis which leads to Brownian motion. This is a remarkable subject for which a large number of specific formulae are known, allowing detailed calculations. They then became aware that other interesting objects also possessed the centre of gravity property from which they had started[10]. The first mathematician to specifically study the processes having the centre of gravity property was Paul Lévy, who generalised Kolmogorov's strong law of large numbers[11]. Jean Ville[12] introduced the term *martingale* used to designate such a process. This name rapidly caught on and replaced the expression of *set of linked variables*, which was used by Paul Lévy (or *processes that have the E property* employed by J.L. Doob[13]).

The Martingales of Mathematicians

This short simple name proved to be useful, and this idea from 1940 to today has experienced an impressive ascent through the ranks of mathematics, becoming the key to some of the most powerful calculations and inequalities, and the heart of a new theory of integration. It was given a big boost by Doob who discovered the first notable inequalities and observed that the centre of gravity property was not only iterative but could be shared in a random way (*Doob's stopping theorem*). If a gambler decides to frequent a casino for at most one month, or perhaps less because of the *random hazards of the game*, the expectation of gain is always zero. In other words, over a determined and pre-fixed duration, all possible strategies give a balanced result between the probable gains and losses. The gamblers follow the dream of their martingale secret weapon, but do not take into account that they can only play for a very limited time.

---

[10] In modern terms, Louis Bachelier was not correct to talk about mathematical martingales (that is to say to talk about the properties of iteration of the centres of gravity) dissociated that of the Markov processes, since they depend on independent increases. Hence he proceeds directly to the Chapman-Kolmogorov property, which, said Smoluchovski, characterises the Markov processes. Cf. J.L. Doob, *Stochastic processes*, Wiley, 1953.

[11] A. Kolmogorov, "Sur la loi forte des grands nombres", Comptes Rendus de l'Académie des Sciences, Paris, 1930.

[12] J. Ville, *Etude critique de la notion de collectif*, Gauthier-Villars, 1939.

[13] J.L. Doob, "Regularity properties of certain families of chance variables", Transactions of the American Mathematical Society, 1940.

As history relates, mathematicians ironically used the term *martingales* to designate high-risk situations where the gambler's secret weapon does not work. When tossing a coin in a casino, the gambler's winnings, *whatever his strategies*, are mathematical martingales. The *martingale* is therefore at the same time one thing and its opposite, a property that it shares with many things that dominate our subconscious. It is on the one hand the pleasure of the prospect of omnipotence, and on the other the display of all of the strategies in a way that destroys them and shows their faults. By adopting this term, mathematicians have underlined that science easily unveils the illusions of gamblers whose behaviour is unscientific. They have also clearly affirmed that gambling has the right to be cited as a scientific subject. This is relatively surprising if we recall the moral judgements aimed at money games and the anathema delivered by the poet Charles Péguy: *"One does not play!"*. Not only does science talk of gambling, but finds productive ideas within it. In fact, science is the best gambler!

Mathematical Development outside the Finance Sector

Mathematics Ahead of its Applications

Despite Bachelier, the development and progress of mathematics (which was considerable during the 20th Century) remained relatively disjoint from finance until a major connection was made in the 1970s. Economics as a discipline had certainly made increasing use of mathematics, as much at the model level as in the treatment of data with a view to economic forecasting. But current finance – stock exchange and actuarial techniques – involved little more than a set of log tables. As recurs throughout the history of science, the concepts useful for the management of what we call *new financial products*, or *derivatives*, provided by mathematics, had been developed for the study of completely different phenomenon: thermal currents, Brownian motion, spectral analysis of signals, noise filtration, etc. More precisely, the mathematical concepts which changed finance, and that are now used daily in Chicago, Paris, London or Singapore, and which will be described in a later chapter, were developed within *Pure Mathematics*, that is to say from investigations within mathematics itself.

Before the Second World War, the probabilistic methods that were used in applications, economic forecasting, the estimation of natural risks, treatment of signals and in the design of bridges and tunnels, were based on the theory of stationary random processes developed by Norbert Wiener. For a long time this simple theory was the only theory taught to engineers. After the war however, a more powerful theory was developed, the *theory of stochastic calculus*, that dealt with situations governed by non-linear equations. This rapidly succeeded the old theory in certain fields, signalling in particular, where it provided an algorithm for noise dissipation and *Kalman filtering*, which led to considerable progress in the performance of guiding or control systems. In other fields the changes in practice were much slower. Even though the theory of stochastic calculus was particularly well adapted to the laying out of construction work, subject to chance conditions (wind, swell, etc) because the equations that govern mechanical deformation are heavily non-linear, it was still the Wiener theory that was applied with linear approximations to design bridges like the Pont de Normandie in France, the Severn Bridge in Britain and the Golden Gate Bridge in the United States. However, in civil engineering, practitioners cautiously welcomed these innovative methods that allowed scientific progress. There are standards and technical rules that represent knowledge, on which are based the collective responsibilities and the *state of the art*, which develop slowly. In finance, by contrast, as soon as researchers (F. Black and M. Scholes) discovered the usefulness of the theory of stochastic calculus, these new ideas were rapidly welcomed throughout the world. Practitioners appropriated and perfected them as though they were exactly what they needed.

The history of science provides us with many examples where mathematics seems to have been developed in a premonitory fashion. At the beginning of the 20th century science had at its disposition the intricate tensors of the preceding century, and quantum mechanics had Hilbert spaces, that had been conceived as generalisations of spaces of several dimensions to resolve the problems of partial differential equations by the development of a series of functions. These are not exceptional coincidences, nor fortuitous phenomena. Using finance as an example gives us the chance to emphasise here an observation in the philosophy of science. Because of the difficulties relating to international agreement, globalisation, credit restrictions, etc., one would like science to remain *closely linked* to that which is economically profitable, and not to become lost in vague investigations. This is what successive politicians announce so definitively on their election, regardless of their political leaning, by devoting themselves to reforms that will make research more useful. But is it by placing all available expertise to work on concrete problems that they are resolved? These problems are often of unsurpassable difficulty without some new ideas. One can set a large group with extensive IT capabilities to work without finding the optimal management of a fleet of trucks distributing freight between different towns. Essentially, mathematics provides ideas and concepts in order to read and understand the inextricable, and then the economic players use this knowledge. The same thing occurs in other sciences; physicists and chemists invent new materials with unusual properties, that have often been long awaited in order to resolve some problem. This is how research works[1].

Science proposes and industry disposes.

We discovered liquid crystals before knowing that they could be used to make chromatic thermometers or watch displays. As for stochastic mathematics, not only has this allowed the financial sector to achieve considerable profits that, related to the number of mathematicians, places the activity in this discipline amongst those with high added-value[2], but it has also improved telecommunications, rocket guidance, the design of off-shore platforms that are subject to heavy swells, robotics, ultrasound and so on.

I will quickly run through the main routes of historical development, that enabled mathematicians to provide the finance sector during the 1970s with efficient and new tools. This will temporarily distance us from our current objective, taking us to a concept which has truly fascinated mathematicians throughout the 20th Century: Brownian motion.

Brownian Motion

The movement of a particle of pollen in the air, discovered by botanist Brown, was modelled by Einstein and Smoluchovski[3] by assuming that the displacement of a particle between two instances t_1 and t_2 is independent of its earlier position and that its law of probability only depends on $t_2 - t_1$. By infinitesimal reasoning similar to that used by Bachelier, they established the basic formulae that govern the phenomenon[4].

Jean Perrin and Léon Brillouin subjected these to detailed experimentation[5].

---

[1] Certainly in return science is itself influenced by social and economic conditions, as one has known at least since Marx, and the sociology of sciences underlines it again today. This gives meaning to political incentives.

[2] It is difficult to know the amount of profits realised, but one can get a good idea from the fact that the derivatives in current use, for the management of which the concepts of stochastic mathematics are used if not indispensable, represent a value in excess of twenty billion dollars.

[3] A. Einstein, Annalen der Physik, 17, 1905, and 19, 1906. M. Smoluchovski, Bulletin de l'Académie des Sciences de Cracovie, 1906, and Annalen der Physik 21, 1906. For a history of Brownian motion, see J.P. Kahane, J. P., "Le mouvement brownien", Actes du Colloque J. Dieudonné, Nice, 1996.

[4] They made the same error as Bachelier concerning the Gaussian hypothesis, but it was found later that this error of reasoning disappears if the particle does not jump (continuous path), which is reasonable here in physics but not, it goes without

Brownian motion Position of a Brownian particle in the vertical direction as a function of time.

It was principally Norbert Wiener[6] in the 1920s who, following the work of the physicists, developed the theory of Brownian motion. A new intellectual adventure was set into motion. The subject was found to be intimately linked to central parts of functional analysis, notably the theory of harmonic functions, the Laplacian operator, and so on, and particularly to the theory of integration, that had delighted theoreticians since ancient Greek and Roman times and was definitively thought to be crowned by the Lebesgue integral. Wiener succeeded in adding a new chapter by showing that Brownian motion allowed the definition of a new integral, that could be used to make a Fourier transformation of random signals, as would normally be done to analyse frequencies, but with specific properties that opened the way simultaneously to simple and useful applications, as well as the solution of profound mathematical problems.

---

saying, in finance.

[5] This verification is related in the book of Perrin [72], where he wrote *"One cannot draw a tangent, even in an approximate fashion, at each point of the path and this is because where it is really natural to think of continuous function without derivatives that the mathematicians have imagined, while the nature suggests functions with derivatives."* Such a continuous function without a derivative anywhere was explicitly described by the Czech mathematician Bolzano a century before.

[6] N. Wiener, "Differential space", Journal of Mathematical Physics, 1923.

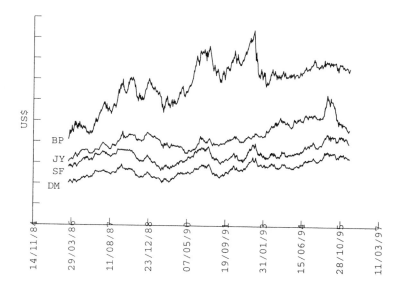

Long-term development (10 years) of the price of the dollar against the pound sterling (BP), Japanese Yen (JY), Swiss Franc (SF) and Deutschmark (DM).

The Stochastic Integral as the Speculator's Gain

The stochastic integral became essential in finance because of its generalisation by the Japanese probabilist Kiyosi Itô. It is therefore important that we look more closely into the idea of the integral.

The Concept of the Integral since Archimedes

The integral is an engineer's idea, simple and easy to understand. To calculate the volume of a cone resting on its base, we divide the cone into fine horizontal slices. Each slice has a volume approximately equal to that of an upright cylinder of the same thickness and radius and is therefore easy to calculate. The sum of these volumes is thus equal to the volume of the cone. The point at which Archimedes, a great engineer himself, made his entry into mathematics was when he used this method of slices with exact precision, obtaining by faultless deduction based on internal and external approximations the beautiful results of integration such as the equality between the surface area of a sphere and the lateral face of the circumscribed cylinder, or the fact that the ratio of the volume of this cylinder to that of the sphere is equal to 3:2.

The integral is a very natural limit object. We can estimate the length of a curve by imagining it rolling along a straight line without slipping, or the area of a region of the plane by filling it with small squares. Obtaining such results by the addition of small rectilinear or planar elements following the method of Archimedes is a legitimate exercise, that has persisted in various forms throughout the history of mathematics. Mathematicians from Descartes and Leibniz through the 18th Century to Euler developed links between integral calculus and differential calculus, where integration appears as the inverse operation to differentiation. The idea concerning the sum of a large number of small quantities, when the magnitude is explicitly governed by an increasing function, was taken up again by Cauchy at the start of the 19th Century so he could extend these results on integrals to the complex plane, and then by Riemann who was the first to completely describe the class of functions for which the integral can be calculated using the method of Archimedes.

A distinct step forward in perfecting this tool was made by Henri Lebesgue in his thesis (1901), and in his later works,

> ... that would make the Lebesgue integral and its generalisations amongst the most living and most used instruments in modern analysis

as stated by Jean Dieudonné[7].

In fact, Lebesgue abandoned the Archimedean slice method for another method whose usefulness comes from the fact that not only does it enable the accurate calculation of the integral for a class of functions larger than that of Riemann, but also that the *integral* tool obtained in this way manifests itself with a greater simplicity because simple theorems can be established that had not been valid previously. As well as Lebesgue, the Dutch professor Stieltjes from the University of Toulouse must also be mentioned, along with Emile Borel, Johann Radon and Arnaud Denjoy.

This integral adapts well to abstract spaces such as those that are used to describe events of probability theory. In 1933 this enabled Kolmogorov to give strict foundations to probability theory that are still in use today, and as a result of which it became possible to talk of more subtle ideas like random functions, or what actually means the same thing, stochastic processes[8].

In particular it was this new integral, or *measure* as it has been called since Lebesgue, that Wiener used in 1923 to describe the law of probability for Brownian motion and with which he accurately established a large number of its

---

[7] Jean Dieudonné, "Intégration et mesure", in *Abrégé d'Histoire des mathématiques*, 1700-1900, edited by J. Dieudonné, Hermann, 1978.

[8] A.N. Kolmogorov, "Théorie générale de la mesure et calcul des probabilités" (in Russian), Troudy komm. acad. sect. matem., 1, pp. 8–21, 1929, and "Grundbegriffe der Wahrscheinlichkeitsrechnung", Erg. Mat. 2, no. 3, 1933.

surprising properties: non-differentiability (the trajectories not differentiable at any point), infinite total variation (over any finite period of time the trajectory is of infinite length), finite quadratic variation, etc.

The Stochastic Integral and Itô Calculus

As a result of his detailed study of Brownian motion, Norbert Wiener discovered a new integral. This was named the *stochastic integral* because of its random nature, and was defined by a new global process taking advantage of the particular properties of Brownian motion.

As we said above, these developments bore no connection to the finance sector. Nevertheless, following the thinking of Bachelier, the financial interpretation of Brownian motion as the price of an asset on the stock market allows us to understand very simply the Wiener integral.

If at the beginning of the quotation month you buy one unit of asset and at the end of the month you sell it, your *profit* will be exactly the difference between the price at the end of the month and the price at the start of the month. If instead of one unit you buy several and sell some part the second day, buy again the third, and so on, your investment is governed by your chosen daily stake function. Your *profit* (algebraic – plus or minus – because this can also be a loss) will be the Wiener integral of the given function. Therefore, as the price is random, we see that the Wiener integral of a function is a random variable.

However, when an operator speculates, he does not normally choose his daily stake function at the beginning of the month – he takes account of the development of the price itself. In this way he will attempt to buy more when the price is low, and sell when it is high. This leads us to the famous *Itô integral*, the last stage of our journey into the theory of integration. In 1944 the Japanese mathematician Kiyosi Itô realised that it was possible to extend the Wiener integral to include functions that are themselves random, provided that these are not *anticipating* over the Brownian motion[9].

In our example, the speculator's stake may be random, but at a given instant, he can only assess the risks from what has happened up to the instant in question. The speculator cannot make use of the future price because, although he could estimate it, it is impossible to know the precise numerical value for it in the future.

The works of Itô and those who followed putting the finishing touches to the stochastic integral, were still without any real link to finance. They were motivated mainly by pure mathematics (probabilistic interpretation of the theory

---

[9] K. Itô, "Stochastic Integral", Proc. Imperial Academy Tokyo, 1944.

of potential, elliptic partial derivative equations, and the representation of the process of diffusion using Brownian motion). But following the lines of thought of Bachelier, we observe that the Itô integral provides us with the (algebraic) profit of the speculator whatever his stake strategy. The hypothesis drawn up by Bachelier states that the expected value of this profit is zero. This brings us back to the starting point of Bachelier, according to whom the expected value of a gambler's winnings is zero.

Because the *centre of gravity property* is satisfied by centred Brownian motion, mathematicians soon realised that the principles of the Itô integral were also applicable to any process having this property, that is to say that Brownian motion could be replaced by a *martingale*. We therefore move away from the Itô integral with respect to Brownian motion, to the Itô integral with respect to a martingale. We already know that the gambler's winnings in a casino have an expected value of zero, that enables mathematicians to imagine the games being *continuous*: it is as if at each instant the stake is governed by a lever which the gambler uses continuously. This type of portfolio management barely presents any difficulties with modern information technology.

It only remains to combine the *Itô integral* with respect to a martingale and the Lebesgue integral and we obtain what is today called the stochastic integral. This is a general, flexible mathematical tool which has links with a completely new differential calculus, different from that used by Leibniz and Newton. Its core formula, the *Itô formula*, that allows changes in variables, is widely used today.

After all of this mathematical work, we arrive at an extremely simple situation in finance: *whatever modelling of the price of the asset* (this may be represented by a martingale such as centred Brownian motion, or by a more general random process that does not have the centre of gravity property), *the algebraic profit that the speculator will gain from such a strategy is the stochastic integral of the stake function with respect to the price of the asset.* This important result is evidently subject to several mathematical hypotheses that will not be explained here, as they are widespread and generalist. This fact was to play a pivotal role in the 1970s when mathematics reunited with finance. These rediscoveries will be covered in subsequent chapters.

In order to give our discussion the strongest possible basis, it is first necessary to examine how a particular way of thinking can lead to setting the price of an object between a vendor and a buyer.

Mathematical Expectation and Best Rational Estimation

More often than might be imagined and for a long time, certain goods have been linked in a formalisable mathematical manner to other goods, that themselves are recognised as having a value. A very old example is given to us by the Mesopotamians. They practiced triangulation to measure the area of fields whose shape was not rectangular. Very early on commodities were sold by weight, and people transported them over distances, while in the building industry quotations were drawn up based on a schedule of quantities that provided an objective basis for discussion. When science uncovered the concept of energy at the beginning of the 19th Century, allowing the comparison of machine work, this was seen as a *mechanical currency* according to the expression of the engineer Navier, who in 1819 wrote:

> *The comparison of diverse machines, for both merchant and capitalist, is naturally carried out according to the quantity of work that they execute, and the price of this work. To estimate the respective values of two wheat mills for example, you need to examine what quantity of flour each can mill in a year; and to compare a wheat mill with a saw mill, estimate the value of the first according to the quantity of flour milled annually and the cost of the milling, and the value of the second according to the quantity of wood that it will saw up in the same period and the price of the sawing. You can limit yourself to considering the machines and the work that they carry out as long as it is only about buying or exchange between them, and where the product is known; but there are many cases where this is insufficient.*

> *Suppose a man owns a wheat mill, and, by making changes to its mechanism, would like to convert it into a sawmill. He can only judge the advantage or disadvantage of this operation if he knows how to evaluate, in relation to the quantity of flour produced by his mill, the quantity of wood it would yield. However this evaluation is absolutely impossible unless you have found a common measure for these two jobs so different in their nature. This example is sufficient to show the necessity of establishing a type of mechanical currency, if one can explain it in this way, with which the quantities of work employed to carry out all types of manufacturing can be estimated[10].*

---

[10] This mechanical money will be work in the mechanical sense of the term. B.F. Bélidor, Architecture hydraulique ou l'art de construire, d'élever, et de menager les eaux pour différents besoins de la vie, new edition with notes and additions by C. Navier (1785–1836), F. Didot, Paris, 1819. See also the article of

It seems totally natural today that the concept of energy should form the basis of our electricity bill. Clearly economic science contributed to this approach by introducing a certain way of thinking on the evaluation of the prices of objects and of services. It is reached relatively easily depending on the case in question. Today where pollution is concerned this raises a difficult but perhaps definitive problem. Evaluations are particularly delicate when risk enters the calculations. The development of marine insurance in the 18th century relied on the calculation of probabilities, that had been motivated by the vogue for gambling games in the salons where the rich landowners mixed with entrepreneurs and prudent businessmen. The calculation of probability made insurance a serious game.

An insurance contract is by its nature a very curious object. Its execution always seems to reveal a priori a conflict of interests. If the cargo of an insured vessel arrives safely in port, its charter company thinks the insurance premium too costly. If the vessel is shipwrecked, then the insurance company thinks the premium insufficient. So what is the object exchanged for which a price must be agreed? The object of the exchange is the risk. The charter company gets rid of it, and gives it up to the insurance company. For that they have to pay a premium. The amount of premium is fundamentally based on an evaluation of the risk. It is about contemplating the respective probabilities of the different possible events and, for each of those the amount in damages, that is to say the stakes.

But risk is an issue that cannot be easily analysed without the help of mathematics[11]. To make a decision in an atmosphere of uncertainty, different random situations need to be compared. Pascal and Fermat were the first to provide a criterion for this effect: the *mathematical expected value*. This is not the only one, far from it; there are other criteria, starting with moments of higher order, or the economists' expected utility, and so on. But it is still against this that the others have to be compared since it adapts to complex random situations in a coherent way.

If when you take the Metro[12], you want to minimise the ground you have to cover to get to the exit without knowing if, at your destination station, the exit is at the front or rear of the train, you automatically adopt the criterion of mathematical expected value by moving to the middle of the train. Unless you estimate two chances in three that the exit will be at the front, in which

---

K. Chatzis: "Economie, machines et mécanique rationnelle: la naissance du concept de travail chez les ingénieurs-savants français entre 1819 et 1829," Annales des Ponts & Chaussées, no. 82, 1997.

[11] Many very simple questions are difficult to treat without mathematics, for example this one: my cousin has two children, one is a girl; what is the probability that the other is a boy? Answer 2/3.

[12] The Metro is the name of the Paris underground system. (Translator)

case it is necessary, according to expected value criterion, to stand two thirds of the train's length toward the front. These calculations allow you to reduce the quadratic risk, also called the *variance* of your voyage on foot[13].

If in order to get from your home to work, other choices present themselves in a random way, and if you adopt the expected value criteria for each of these in succession, they will be consistent and you will reduce your overall quadratic risk on the journey time. The mathematical expected value, and *conditional expected value*, amounts to choosing from amongst the possible distances, the one which reduces the quadratic risk (variance).

Saying that the current price is the value of the price today, the one that makes the average of the squares of the variations between today's price and the price tomorrow as small as possible is the same as saying, like Bachelier, that the current price is the expected value of tomorrow's price. An analogous relation is well-known in mechanics: the centre of gravity of a rod, the point about which it rests in equilibrium, is also the point about which it has the smallest moment of inertia.

To choose a future position according to the expected value criteria (or conditional expected value) is generally considered the most sensible decision. Although it is not the only one available, it comes down to reducing the risk – in the sense of the least-squares – meaning the variance – and that gives it both consistency and simplicity.

Precisely because the expected value is a best rational estimation in this sense, and because a martingale takes as a value at each instant the expected value of its future values, some stock market prices are martingales. P.A. Samuelson demonstrated this in 1965 regarding *futures*: the levels where futures are set, that is the daily quotes for the purchase of an asset at a *fixed* later date, form a martingale if the prices represent the expected value. These works and others earned him the Nobel Prize in 1970. They introduced the world of finance to a large-scale shift towards the new tools that mathematicians had developed, notably the stochastic integral and the Itô formula.

Can Samuelson's result be seen in price trajectories? How do you recognise that a function has a martingale trajectory? As we only use a finite interval of time, there is no certain answer. However martingale trajectories cannot be completely arbitrary. If they are continuous (which is a current hypothesis, although the opposite hypothesis is not without interest) they inevitably resemble (locally) Brownian motion, they have the same steep appearance and the same degree of irregularity. This is also observed in prices, not only those for *futures*. Whether prices are martingales or not is a question that will continue to be

---

[13] If, instead of the random variable X, you choose the number a, the quadratic risk $E[(X - a)^2]$ is minimal when $a = E[X]$. The quadratic risk is then equal to the variance $E[(X - E(X))^2]$.

discussed for a long time to come. When this is the case, we sometimes say that the market is *efficient*[14], and that this goes back to an economic thinking based on the behaviour of agents where each optimises their advantage, the risks being calculated by variances of uncertain quantities.

We repeat that even if the market is not efficient and the rate of an asset is not a martingale, when you speculate on an asset using a strategy, however clever, your profit is given by a stochastic integral with respect to the price of the asset[15].

---

[14] We will discuss this concept in detail later.
[15] The questions of updating have been omitted here for simplicity.

Part II

Option Hedging: an Epistemological Rupture

The work of Bachelier linking the prices on the stock market to a mathematical model of risk, while interesting in itself, remained unknown to financiers for more than half a century. But as we have seen it was extended by the mathematical study of Brownian motion and of martingales up to the elucidation of the stochastic integral of Itô.

I want to show in this Part how, and for what particular reason, this lengthy mathematical work from more than seventy years earlier came to be applied to finance and to furnish it with the methods and concepts that today underlie the operation of the financial markets.

I will talk about options, financial products with a long history, that are exchanged on options markets, and I will examine the question of the price for these contracts. Their pricing is made on the basis of an argument of non-arbitrage, which takes account of the prices that are provided continuously by the market and has allowed the discovery of the *principle of hedging* for the management of options.

After this I will talk about the implications of this way of thinking by comparing it with the best estimate in the sense of least-squares, and I will develop some consequences of this on the epistemological plan.

4
The Unexpected Connection

A *futures contract* is the conclusion of a transaction that has a settlement in the future under conditions set today. An *option* is a contingent futures contract, that is to say the conclusion today of a transaction that could develop in various ways according to various outcomes on the result of which the transaction depends. One such option, the simplest, is the option to buy or a *call*. This is a contract between say a bank and a client giving the client the right, but not the obligation, to buy at date T (perhaps 3 or 6 months), specific shares or a currency at an agreed price K, even if the price of these shares or currency is greater than K at the date T. The possession of an option is clearly an advantage to the client. The client may be a business where the expenses and income are seasonal and delayed. For example, a French business that buys oranges in winter in dollars and sells orange juice in summer in euros will see that the annual balance sheet is extremely dependent on the fluctuation of the dollar in relation to the euro between winter and summer and between summer and the following winter[1]. In the summer it could take up an option to buy in dollars in six months time which would give it a sort of insurance against the variation of prices.

The question that we are going to discuss is that of the price that the bank

---

[1] For example Louis Gallois, then CEO of Aérospatiale, stated in Le Monde of 22/09/1995: *"In the first quarter [of 1995] the net loss of Aérospatiale was 105 million francs, compared with a loss of 333 millions in the first quarter of 1994. If the exchange rate with the dollar had been the same one year later, the group would have benefited by 655 million francs. That is to say that by itself the dollar had had a negative effect of 760 million francs over the corresponding quarters".*

must charge for such an option, or more precisely, the method that is followed in order to set the price and its economic justification.

Options are traded today over the counter (OTC), which is a market in the current sense of the word between the banks and their clients, as well as on the *organised markets* where the interplay between the supply and demand is continuously maintained and provides a quotation for buying and selling with low transaction costs. Such markets have existed for a long time for traditional assets (shares, raw materials, currency, bonds) as well as for options and other derivatives that are futures contracts on the traditional assets, or on negotiable indices[2]. The development of these organised markets for derivatives has made a considerable expansion during the last thirty years or so.

Organised Markets

The story of organised markets goes back to the middle of the nineteenth century with the creation of the *Chicago Board of Trade* (1848) where cereal producers could delay the delivery of the harvest on a certain date by agreeing a price beforehand. For them it was thus possible to be free from the risk of falling prices. For them it was sufficient to find a buyer who, on the contrary, anticipated an increase in prices. An agreement could then be made in return for payment, by one side or the other, of a compensation (called the premium of the trade). In 1874 the *Chicago Mercantile Exchange* opened and remains its main competitor[3]. Now the markets agree to futures options that are themselves quoted and exchanged just as for ordinary assets.

France did not wish to be outdone however, and the Bourse de Commerce de Paris opened for futures contracts in 1884. This market closed in 1939, then a new one opened with trade only in white sugar to close finally in 1976.

Options on currency appeared the first time in 1972 in Chicago, a little after the decision to abandon the Gold Standard in August 1971, and the devaluation of the dollar four months later, that made apparent the risks attached to exchange rates. Options on interest rates appeared in 1973[4].

Between 1978 and 1990 markets opened for derivatives in various financial

---

[2] Baskets of assets such as the CAC40 index in France, created in June 1988 that serve to support contracts quoted on the MATIF and MONEP, see "MATIF" in the Glossary, page 138.

[3] Together in 1992 they exchanged four times more contracts than that of the third-place world market, namely London.

[4] For a complete panoramic history, see C. Hersent and Y. Simon, *Marchés à terme et options dans le monde*, Dalloz, 1989. For more details and anecdotes, see Bernstein [11].

centres throughout the world: in New York in 1979, London (LIFFE) in 1982, Singapore (SIMEX) in 1984, Tokyo (TIFFE) in 1985, Paris (MATIF in 1986 and MONEP in 1987) and in Frankfurt (DTB) in 1990[5].

After the 1970s, the derivatives markets developed rapidly. The prices of underlyings (raw materials, assets, currencies) *resemble* Brownian motions, but we do not know whether they are martingales or not.

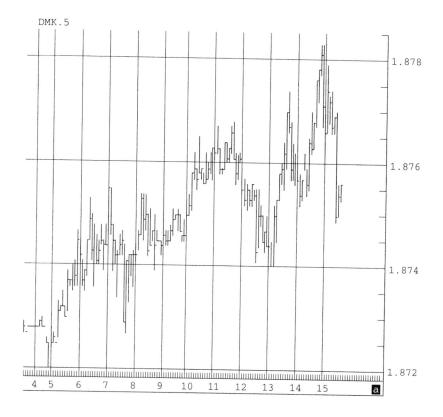

To compare price variations with Brownian motion is quite natural. These processes can be investigated with a particular form of integral calculus, the Itô calculus. This diagram shows the movement of the price of the dollar in marks in Paris between 5:00 and 15:00 on 15th September 1988. Each vertical bar represents the amplitude of variation of price during the last 5 minutes.

---

[5] Other countries have opened both classical and derivatives financial markets, notably those needing foreign investment. These markets are called *emergents*, and are often more volatile than others. The insurance aspect of derivatives plays a very important role here.

We are going to establish an extremely interesting argument with the promi-
nent place being the reference to the instantaneous price given by the market
and the way it exploits properties of Brownian motion. It appeared for the first
time in this particular form in an article by F. Black and M. Scholes [12] in
1973.

Hedging Portfolios and Non-arbitrage

What allows the setting of the price for an option is the concept of a *hedging
portfolio*. Suppose that the option is about estimating in three months the price
of the Japanese yen. All the speculations we can envisage about the yen (which
is here the underlying of the option) can be described by a portfolio comprising
only euros and yens, in algebraic quantities (positive or negative), which is
initially empty, and will after three months contain only euros. If at a certain
time we buy yens, they come into our portfolio but their cost in euros is written
down with a minus sign. If we sell some yens at another time at a different price,
the amount of euros is raised, and finally at the end of the three months we
sell all the remaining yens so we only have euros in the portfolio. What do
we obtain by managing the portfolio over these three months by buying and
selling yens? This is certainly not a balanced operation, at least not in general.
It produces a profit or a loss, a variable result in each case.

For the reasons exposed in the previous chapter, this random result is exactly the value of the stochastic integral of the quantity of yens held in the
portfolio with respect to the price of the yen.

To find the correct price of an option, that is to say of a futures transaction,
the principle is as follows: *if a random variable can be written as the sum of a
constant k and a stochastic integral with respect to the price, then the benefit of
disposing of this random variable at its maturity has a fair price and this price
is k (including the premium).*

If the bank does not set the price exactly at k, then either the bank or its
client will make an *arbitrage* and realise a profit without risk. How? Suppose
that the bank sells a futures option more expensively, at say $k + h$. In making
a portfolio and managing it by buying and selling in a way that it realises the
stochastic integral in question, it will be the same as realising the contract, and
it will gain h. In the same way, if the price is set at $k - h$ it is the client who,
with the constitution of the analogous portfolio, will realise a profit amounting
to h.

The reason is the *absence of arbitrage*. It carries a self-imposed logic. Because if for some reason it was not applied, then profits without risk could be

realised, and the agents would apply it again very quickly.

If the correct price thus determined – and we will see later how one can obtain a numerical value – is accepted by the bank and by its client, how will the bank be able to furnish exactly the desired amount to the client? By constituting the very portfolio we have been talking about, called the *hedging portfolio* or *simulating portfolio* which, if it is managed according the correct strategy, removes risk from the operation.

We are assuming here that the random variable which is the option can be written as a constant plus an integral with respect to the price of the underlying. This integral is the result of a speculation on the price of the underlying, that is to say the result of buying and selling in a portfolio containing suitable quantities of yens. Thus one can *simulate the option up to a constant by this portfolio*. The constant plus the portfolio have exactly the same characteristics, the same financial use, at each time, as the option.

To make this program work concretely one needs certain hypotheses on the price of the underlying asset. In the Black-Scholes model – which is still the most used model – the price is described by a single parameter, the *volatility*, that describes its agitation. This model also has the property that *every* random variable that one can envisage concerning the underlying can be written as the sum of a constant and a stochastic integral. One says that in this model *the market is complete*[6]. All the options can then be made the object of a hedge by a simulating portfolio; the calculations are explicit for the common options and there are numerical algorithms (Monte Carlo methods for instance) for other cases. They indicate the amount of the premium and the composition of the hedging portfolio at each instant.

This particular model, which was published by Black and Scholes, remains much used because of its simplicity[7]. It presents the possibility of constituting a simulating portfolio that represents the epistemological rupture because it was so very different, and we will return to it later, from that given in the works of Bachelier. A substantial theory of mathematical finance developed rapidly that gives all its generality to the principle of hedging and puts to work the

---

[6] Many other models where the price is continuous have also this property. But except for very particular cases, the models where the price of the assets presents jumps do not have this property, and an exact hedging is not generally possible and so it remains risky for the bank. One can always find a hedge that minimises the risk, cf. H. Föllmer and D. Sondermann, "Hedging of non redundant contingent claim", 1986, and N. Bouleau and D. Lamberton, "Residual risks and hedging strategies in Markovian markets", Stochastic Processes and Applications 33, 1989.

[7] And its robustness cf. N. El Karoui and M. Jeanblanc-Picqué, *On the robustness of Black-Scholes formula*, Laboratoire de Probabilités, Université Paris VI, 1994, and N. El Karoui, M. Jeanblanc-Picqué and St. Shreve, "Robustness of Black and Scholes formula", Mathematical Finance, 8, 1998.

progress made in probability theory since Bachelier[8].

In developing this, the derivatives markets make more and more appeals to applied mathematics not only with respect to options but also for *firm products* (*futures, swaps*, etc.), which are the object of modelling of interest rates. As for those concerning options, the principle of hedging applies itself on one hand to European options – specific terminology that denotes a contract carrying a contingent transaction but for which the date is fixed – such as options to buy (calls), or to sell (puts) or more sophisticated (*straddle, strangle, butterfly, digital options*, etc.) and on the other hand to the corresponding American options – which can be exercised at any time up to the maturity date, that make even more use of the stochastic calculus of Itô and partial differential equations and that also make an appeal to variational inequalities.

Every new product conceived (*caps, floors, barrier options, structured products* etc.), before being put into circulation, is the object of a mathematical study to establish a non-arbitrage price and a method of hedging. This is one of the reasons why the banks need to recruit mathematicians, a topic to which we will return.

---

[8] On options and organised markets, one can consult: Arnould [4], Bourguinat [23], Dana and JeanBlanc-Picqué [29], Duffie [35], Elliott and Kopp [41], Gibson [45], Huang and Litzenberger [50], Jacquillat and Lasry [54], Lamberton and Lapeyre, [60], Malliaris and Brock [62], Merton [65].

5
A Different Approach

The Principle of Hedging and Best Rational Estimation

The possibility of hedging is essential for a product to be spread. When the bank manages a hedging portfolio using the Black-Scholes formula, it is not on average price over its time scale that will determine the amount required at maturity, it is the price at each instant of time whatever the development of risk. Up to the errors of the approximating models, that from experience are sufficiently small for products where there are plenty of transactions, products are said to be *in the money*.

So for the industrial producer of orange juice, for example, the option constitutes an *insurance*. However, it is not managed by the bank as it would be by an insurance company, by sharing out the risk over several assets and hoping to regain it. No, the bank balances its management *for each option*. That does not prevent it from wishing to diversify its contracts and its transactions in order to diminish secondary risks, but the expansion of derivatives and of their markets is built on this particular management (called, for technical reasons, delta neutral management)[1] for which one of the requirements is the need for many transactions on the same asset. This explains the enormous quantities of transactions in currency[2]. This type of management allows the avoidance of risks. It is interesting to note that it makes a direct and systematic taxation on

[1] See "Management of options" in the Glossary, page 136.
[2] See the conclusion of Chapter 6.

transactions impossible without putting into jeopardy the use of derivatives or introducing new serious uncertainties[3].

The historical importance of the evaluation by hedging can be clearly seen if one compares it to the natural method of anticipation in the sense of least-squares in the spirit of Bachelier.

Consider the situation where the bank sells a European call at three months at exercise price K based on some asset. This means that if in the three months the price of the asset is less than K, then there will be nothing to pay. If, on the other hand, it is much greater than K, the bank will profit by the value of the price minus K. Thus in classical thinking, such as was in use until the 1960s, in order to set the price to sell, the bank must try to determine if the price is going to rise or fall. More precisely, it must try to estimate the *probability law* of the price until its maturity and to calculate the *expectation of its loss*, that will set the price which it is going to propose to the client (with a small margin for expenses, a supplementary margin that it takes as it does for a hedging portfolio). In choosing this expectation it minimises the risk in the sense of least-squares. Consequently, the bank is going to ask its experts to give an evaluation of the probability law in question. Its economists, the *fundamentalists* who study the asset and its economic health, are going to do this work. If these studies concern a share, they will examine the performance indicators of the business, its balance sheet, its cash flows, etc.; if they concern a currency, they will look at public deficits, inflation and the interest rates of the country concerned, and so on, in order to understand the probable development of the fundamental, that is to say the economic role of the underlying.

Several remarks concern this management according to the old way of thinking. Firstly, even supposing that the bank experts are competent and that they are clear about the probability law, the bank will not balance its management on this single transaction. It is over a large number of options sold that the gains and losses compensate each other and through which the profit margins are produced.

Secondly, if the client were to go to another bank, he would be given a different price, because the experts are not the same. Some banks have accumulated considerable quantities of information on their computer systems, and others none. The price proposed by each bank will be an equitable price given the economic information it has at its disposal, it is an *expert opinion*.

---

[3] The reasons are technical but one can say that such taxation would have a similar effect to that of a tax on the wear of brakes on traffic! Apart from which, they would penalise the money market when applied.

The Thinking of the Market

With the new thinking of the 1970s based on the possibility of constituting a hedging portfolio and being able to exchange it as one would an asset with its instantaneous prices, the banker follows the true dynamic of the asset. Moreover, anyone can follow this, without trying to effect a judgement on its development. It does not appeal to its experts, it follows the logic of the market and its management is the same as that of a share in a petrol company, or a currency or the price of cocoa. Further more, this method of management, that from an economic point of view looks like groping in the dark, allows it not only to minimise its risk, but to annul it (we will return to secondary risks later). The logic of hedging is a thinking founded on the available knowledge. And (at least in principle, and we will return to this as well) it is a non-speculation. One does not use projections based on the rise or fall of prices, as was the case in the natural method of prediction in the sense of least-squares, that, even though the bank experts know more than anyone else, is still of a speculative nature. Thus the hedging portfolio is founded on the *opinion of the public domain*, that is to say it is accessible to all.

This discussion needs to be refined. For simplicity I have left out several technical details that have a certain importance, above all when rare events happen[4]. Without detailed analysis, we simply note that each of the two approaches needs the construction of a *mathematical model* for the price of the underlying asset, but that the significance of the models is different. The bank activity is not the same in the two cases. If the bank realises its estimate of the price of an option using the Black-Scholes model, its profit is not bound by its economic competence, but only by the margin that it has added. It is a commercial profit of a traditional type, for which moreover it is in competition with the other banks, and which, like for other commerce, will be sensitive to the effect of innovation, but only during a certain time after the appearance of a new product. The derivatives are themselves quoted on the markets, however, and models other than Black-Scholes sometimes fit better the shape of the prices of options according to their characteristics. There is risk on the choice of model, including risks due to transaction costs, risks specific to products *out of the money*[5] for which the markets are not fluid enough, risk of competition, and so on. This sort of bank activity cannot technically be purely commercial, since what is required is a comprehension of the economy on one hand and the pertinence of mathematical models on the other.

Once these further details have been dealt with, one can say that the man-

---

[4] We have notably, to simplify the discussion, passed silently over the updating that must be taken into account.

[5] See "Out of the money" in the Glossary, page 140.

agement by a hedging portfolio of derivatives has given to the thinking of the markets – where the anticipation of players is expressed as a function of their economic force – a quite new dimension that has important philosophical and economic consequences.

We will look at the latter in a later chapter. For the former let us simply note here that it strikes a blow at a conception of the applied economy which in turn strikes at an objective and universal knowledge.

When, in order to try to clarify thinking on the economy, certain economists, researchers, university or political advisers, present themselves as experts with a good model, or consider that a universally valid model does exist, they consider that the assets of the market have a *fundamental* that expresses the underlying real economy. From their point of view, this is not affected in a chaotic way as is the price, because the objective facts do not change every 30 seconds. They investigate whether the averages can approach this fundamental. If the market is spread, it refers too much to itself; it is unstable, it forms speculative bulls.

On the contrary, the thinking of the market considers that the economic reality is the market price, and that, for a given asset, there is only one market price but different estimates of the fundamental by different experts and by different analysts.

Subjectivity of the Laws of Probability

There is a clear cut difference between the two camps, whose agreement is difficult to imagine. One camp feels that the markets are mad and that they move away or risk moving away from the real economy. The other camp considers that any scientific view of the economy is suspect since it plays with analytic and normative propositions.

There is a real reversal of point of view analogous to that of marxism visa-vis the hegelienne philosophy. The concrete replaces the abstract. We will see several implications later in terms of the instruments of political power. But now it appears to me useful to emphasise that a totally objective vision is not tenable. I would like to say that economic science could give pertinent and objective descriptions to reality, but one cannot include here the laws of probability of the phenomena to come. *The laws of probability are subjective*[6].

In the same way that the physical phenomena are invariant under any change of Galilean axes, the random economic phenomena can be described

---

[6] This observation is evidently well-known by many economists, see for example Kolm [58]; it is the management of new financial products by a hedging portfolio that provides the proof of a new force.

indifferently by every class of probability law, quite different from one another and between which it is impossible to choose. This is close, in the case of modelling prices, to an important theorem of the mathematician Girsanov, that clarifies brilliantly the debate on the efficiency of markets. The question of knowing whether or not the price of a certain asset is a mathematical martingale is often tackled in the economic literature by trying to find reasons in the rational behaviour of players so that an equilibrium of an equitable game like in the pure casino is established. One imagines that this property will be objective. But the theorem of Girsanov[7] says:

> If at a casino you cannot see the roulette wheel, but observe only the gains and losses of the gamblers over a fixed period, it is impossible to know if the game has a bias or not.

To put this more simply, over a succession of 10 or 20 throws, it is not possible to see if the dice is fair or not. To define a concept of efficiency that will be objective, it is certainly necessary to give it a sufficiently general meaning. To suppose, for example, that the yen in dollars and the dollar in yens are martingales is a contradiction because a random variable and its inverse cannot both be martingales. The property of being a martingale depends on the point of view adopted.

In other words, the possibility of making speculative profits on a market, or on the other hand the impossibility of such profits in *expectation*, as is the case at the casino, is a question that, in general, cannot be settled scientifically. We will return in detail to this question of efficiency in the final Part.

---

[7] This theorem brings out an important property of random processes and can be formalised in a very general form. For a formulation sufficiently practical for finance, see D. Lamberton and B. Lapeyre [60].

6
Hedging Risks Thanks to the Market

The Strength of the Arbitrage Non-arbitrage Argument and the Principle of Hedging

We will start our discussion with an example. Since the price of the American dollar in euros is quoted at each instant in time, all the players on the currency market who hold a portfolio in dollars and euros can buy and sell continuously one of the currencies against the other. To sell a certain quantity then to rebuy it is not the same as buying the same amount then selling because the price varies at each instant. One clearly tries to sell dollars when the price is high, and buy when it is low, but the price changes in an unpredictable way. Through this buying and selling, the composition of the portfolio varies. Let S_t denote the price of the dollar at time t. If we start with zero dollars, and if the portfolio contains K_t dollars at time t, in selling all the dollars at time T one realises the amount

$$\int_0^T K_t \, dS_t.$$

This quantity can be positive or negative. We call it the *algebraic gain* from the strategy K_t. It is always expressed as a stochastic integral with respect to the price S_t. It is this integral that we must study. We are the masters of the choice of K_t, provided that K_t does not anticipate (i.e. uses only what has happened up to time t). In return, S_t is given to us by the price. By looking at this one can get explicit examples of portfolios.

If we take $K_t = S_t$, or $K_t = (S_t)^2$, or $K_t = f(S_t)$ or even $K_t = F(S_t, t)$

and so on, what algebraic gains will there be? The answer is that one gets a large variety of random variables – so many that *every* random variable that one can define on the price of the dollar S_t between 0 and a given time T is obtained this way up to a constant.

This is an important mathematical result: for every given random variable[1] H defined on the price of the dollar between 0 and T there is a portfolio K_t such that

$$H = \int_0^T K_t \, dS_t \text{ up to a constant.}$$

Unfortunately, the calculation of K_t given H is not explicit. There is no simple formula and one must use approximation methods.

In particular, if we consider an option on the dollar with maturity date T, there is a portfolio K_t such that the value of the option can be written

$$k + \int_0^T K_t \, dS_t.$$

For example, if we know the exercise price C of a *call*, its value is the maximum of $S_T - C$ and 0. We denote this quantity by $(S_T - C)^+$. There exists a portfolio K_t such that

$$(S_T - C)^+ = k + \int_0^T K_t \, dS_t.$$

Under these conditions, the price of the option (at time 0) is clearly k. Because every other price allows one of the parties arbitrage in making use of the market, since with k euros one can manage a portfolio that at time t contains exactly K_t dollars, one realises the option exactly. Moreover this is what the bank will do once it has received k euros (plus expenses) by selling the option to a client. It will make a portfolio called *a hedge*, which it will manage, through the buying and selling of dollars, containing K_t dollars at time t, realising then at the deadline T what is being asked for.

One can see how much this constitutes an epistemological rupture by recalling how the price was set for the options using the thinking of the 1960s. For an option of three months on an index (a basket of options like the CAC40) the bank would appeal to its experts who would draw up a balance sheet of the technical and financial health of the businesses involved and of assets concerned and would estimate the tendency of the index. From this study, the experts would deduce a law of probability for the value of the index in three months, after they had looked at the average value expected for the futures contract, i.e. the option. In this classic vision there is nothing to do after the

---

[1] The random variable H might for example be of the form $g(S_0, S_{t_1}, \ldots, S_{t_n}, S_T)$ with $0 < t_1 < t_2 < \ldots < t_n < T$. In the following discussion we only consider the case where H has the form $f(S_T)$.

sale of the option during the three months other than to wait and see if it has gained or lost in the end. It will only break even in the long term, over a large number of options (provided the expertise is good).

Clearly, one *cannot* proceed in this way since the price deduced by experts provokes in general an *arbitrage*, i.e. a profit without risk for the bank and for the client.

It is clear that the derivatives markets, i.e. the organised markets for options and other futures assets, would not have been able to make such considerable expansion since the 1970s if they had retained this old way of thinking. There were too many risks for the banks, and in consequence the options could not offer the same service to organisations wishing to hedge against future risks, (and, in particular, exchange rate risks) that might cause their costs to rise.

Today, we find a situation where the traditional stock market, where shares, currency and other classic assets are quoted at each instant, coexists with the derivatives markets where there are quotations for new financial products (options, swaps, futures) that are future contracts on classical assets. In other words, the new products are random variables defined as a function of the trajectory of the classic assets.

In the last Part of this book, we will see that the fact that the derivatives are organised into markets is extremely important and has many political and economic consequences. We simply note for the moment that if one considers an option to buy (a call) at three months on the dollar, the hedging portfolio and the monitoring of the price of the dollar allows us to know the price of this option, or at least its theoretical price. One can then compare this to the price given on the derivatives market where the option itself is quoted. In general, there will be a difference in price, small but significant, which depends on the option. On examining this difference one can see whether the market is anticipating a rise or fall of the dollar in three months. As there are also options of six months or longer, it is also possible to see whether the derivatives markets anticipate a rise in the dollar over three months but a fall over six months. This surveillance is the permanent work of traders. Given the large variety of options and diverse maturity dates, one can see that, thanks to the derivatives markets, the expression of the anticipations of the players is more detailed than on the traditional single stock market.

Since the 1970s, the thinking of the market has marked several points. Firstly, by its preference in favour of the *globalisation* of exchanges, surely a phenomenon of major historic importance, and also through the method of organising the institutions that has induced a sensible shift of economic power. The emergence of derivatives and their quotations on organised markets has played a decisive role there. Because the markets can express many things about the economy, they immediately carry their judgements on developments in the short term, and in the long term, on the correlations, on the risks and so on. *They talk.* And to speak is an expression of power. The question is then on one hand about the freedom of speech, on the other hand about its legitimacy.

But before tackling this big question under ideal conditions, I propose to clarify an uncertainty that mixes aspects of economics, mathematics and ethics and which we can leave no longer: speculation. What is it? Moreover, does it exist?

Mobility of Capital and the Hedging of Options

A large number of authors, journalists and economists regrettably perpetuate a political or philosophical confusion in their interpretation of the important flow generated by the management of derivatives. This error is to be found even among better-informed observers.

I have described to you the principle of management of options by the *hedging portfolio* (in delta neutral if one uses the Black-Scholes model) whereby a new way of thinking emerges, induced by the principle of hedging. This is that the quantities of money and of underlying in the portfolio must be *constantly* adjusted to certain values obtained by calculation, that depend on the instantaneous price (and eventually on the whole trajectory of the price since the start of the options). In practice, the manager is not going to adjust the portfolio every minute, because each of these adjustments is going to incur small transaction costs, that will build up. He will do so when the composition of the portfolio has moved away from what it should be. The number and quantities exchanged to manage the portfolio are very variable as a function of the option and of the market, so while taking account of all the other managed portfolios, a useful talent for a manager to have is the ability to do the least necessary.

The quantity of transactions on the markets has grown to over 1000 billion[2] dollars *each day* (in 1993, quoted from the Bank des Règlements Internationaux), so the total cannot then in any way be related to the mass of international capital searching for the most profitable investments, nor can it be interpreted as a mass of speculative transactions. In effect, these 1000 billion dollars are not available for anyone; for each bank and financial institution, the major part of this amount is committed leaving only a small balance.

Giving the wrong opinion is dangerous for the correct understanding of the problems of international finance. Precisely because they are affected by many things, these problems demand a detached way of thinking, removed from passion and emotion, that must not be transformed into some implacable

---

[2] billion = thousand million = 10^9. (Translator)

power, with the only goal being to hold the attention of the reader or television viewer.

Keep in your mind the order of magnitude: one of the most famous incidents of recent years was the speculation in 1992 against the pound sterling that made George Soros famous, in which he netted a billion dollars.

The general budget for France in 1996 was 1264 billion francs, the budget deficit in summer 1996 was 288 billion and that of the Social Security was 51.6 billion. In May 1997, the French stock market capitalisation represented a total of 3347 billion francs.

Part III

Science and Speculation

We continue our study of financial markets by analysing the activities of traders, who, for the banks and investment organisations, manage the funds of affluent and seasonal business accounts, or draw some of their profits buying or selling assets and new financial products in the markets.

7
A Complex Dynamic

The basic characteristic of these practices is that they are complex. They make use of science for its knowledge that cannot be reduced to savoir-faire or to technique, even if the latter plays an important role. This sort of science and speculation is found in relation, in contact with, what one might call *an intelligence*. But that would be to make a value judgement. We note that there is an ethical tendency that, with the same etymology, moves from *complexity* to *complicity*. What is true is that science and finance play together. This is not a secret understanding for deception, so to talk of a *collusion* would be wrong. What do we say then? There is, let us say, a *wink*, or a *connivance* (from *cum nivere*, meaning *to wink*).

The previous chapter showed that we can formulate the problem in the following way: in the casino, the gambler shows a certain naivety, because probability theory shows that all the martingales that fascinate him are only pipe dreams. To the large casino financier, who is the most naive? The trader who calculates possible profits, the investor who schemes or the mathematician who helps one or the other? The question is a lot less clear.

To tackle this, we cannot accept uncritically – something disliked by certain economists – the description of the role of speculators.

We will start from a meaning more general than financial speculation, the search to realise profits by stock market transactions, and we will gradually focus on certain realities in this vast sector of activity. The thesis posed in this Part is articulated around the two statements that I will first make, then criticise and finally amend. The first is that *speculation is judged useful by most specialists, according to academic criteria*. The fact that the scientists arrive at normative judgements about a practice whose legitimacy is evidently a moral and political problem, is sufficiently provocative to warrant further examination. The second is that *in order to speculate well, it is necessary to do it scientifically.* So speculation has become a field of application for science and a professional outlet for scientists. These two assertions reinforce each other, they form a couple, in the mechanical sense of the word, and generate a dynamic whose consequences play a certain historic role.

The Professionals

To speculate today is a skill that was developed in the United States in the 1970s after the creation of the derived markets, under the name *locals.* In France, this skill appeared in 1989 on the MATIF with the intervention of the Négociateurs Indépendant du Parquet on the market index CAC40, and on the futures trades on interest rates called *notional.* It concerns people who have the status of traders and who are controlled by MATIF S.A. and the banks who register them and to whom they are accountable. Their actions represent around 10% of the global volume.

What is the social function of speculators? Suppose that in Paris, by means of an acceptable organisation of a stock market between the RATP[1] and representatives of travellers[2], metro tickets can be sold each day at a price that varies according to supply and demand. Suppose in addition that the tickets can be used at any time chosen by the buyer up to one month after issue. The RATP – going along the lines started by the SNCF[3] with the system SOCRATE[4] – will make them more expensive when there is a lot of demand, and cheaper when there is less.

If anyone can buy and sell tickets, what will happen? Because of the well-known fact that when it rains people take advantage of the metro, the price will increase on these days, so the speculators will be interested in the weather. Those who make their meteorological predictions most accurately will find the opportunity to make a profit on the buying and selling of tickets. These profits are not linked to the movement of travellers, so where do they come from? Clearly from those who do not speculate in the same way, perhaps those who travel rarely and buy their tickets at the last minute, perhaps those who make

---

[1] RATP is the Paris public transport system, incorporating buses and the Metro. (Translator)

[2] I repeat here in essence my discussion of speculation published in Libération of 23/05/94 under the title "Ethique et finance".

[3] SNCF is the French railway network. (Translator)

[4] An advance train ticket booking system for SNCF. (Translator)

a mistake in their weather predictions, and there will often be many of those.

In what sense does the intervention of speculators affect the development of the price of a metro ticket? When it is fine, the price is low, therefore they buy, which raises the price. Conversely when it rains, the price goes up, so they sell, and the price goes down. The logic of their role tends then to attenuate the market fluctuations that come from these causes that they anticipate[5].

For completeness, we observe that not only do the wise speculators provoke a lessening of the fluctuations of the price, but in addition that many of them take risks. The system of meteorological forecasts being what it is, those who possess information can find themselves wrong and, if they speculate, can experience losses instead of profits. It all depends on the validity of their predictions.

Speculation and Exchange of Risks

The example developed above is given in the spirit of the definition given by the economist Nicholas Kaldor in 1939. According to him, speculation is *the selling (or buying) of merchandise with a view to rebuy (or resell) at a later date, where the motive of such an action is the anticipation of a change of current price and not an advantage resulting from their use or their transformation or transfer from one market to another.*

The speculators who widen the market by their transactions are producers of liquidity. A market needs lots of supply and demand. Each quantity exchanged changes the supply and demand brutally, and the price adjusts badly. Because of their anticipations, the speculators take on some risks that weigh more or less than others. When a farmer chooses to sell his crop in advance because he fears a fall in price, the person who buys it from him generally speculating on its increase. Presumably the farmer sells a little less than the average but he gets a price known in advance. He transfers the risk to the speculator by accepting payment which is a kind of insurance premium.

A similar approach classic for a student of DEUG[6] Economics, would provide evidence that speculation is useful. It deadens the fluctuations of market prices and allows the transfer of risks to those who wish to cover them.

Can we conclude that the dossier labelled *speculation* is complete and can be closed? It is important to remember that our world legitimately creates a more detailed critical thinking on the financial markets as long as their place

---

[5] A condition that their effect on the market price stays weak. If not, among the causes of fluctuation, one must take account the reactions of speculators themselves, and it may be that the speculation is much more amplified in this last type of fluctuation.

[6] DEUG is a French university undergraduate qualification. (Translator)

is central to the operation of the economy. Can one reasonably consider that the allocation of resources is at its best when young people from the suburbs have no future, the disparity between rich and poor increases and the fear of movement of capital leads certain governments to take measures to kill investment, employment, or systems of social protection? Some economists push this further, on one hand by asking whether these threatening evils do not come from obstacles to the free trade of the market (price ceiling, quotas, taxes, protections, government stocks), on the other hand, according to a more Keynesian perspective, evoking the problem of revenue sharing, of speculation, of tax opportunities and so on.

Whatever the cause, today one cannot be content to think that speculators widen the market and take onto themselves the risks that others do not want.

The reality is infinitely more complex. Some economists see much simplicity, equilibria and harmony, while other economists see little. But all search constantly, as statesmen, for ways to combine full employment, stable prices, economic growth and equitable international relations.

Our approach will be descriptive and phenomological. We will first study certain difficulties concerning the analysis of links between chance and risks that are in the background of the market operators. This will allow us to tackle more pertinently the principal modes of speculation and to review this analysis of speculation from a much clearer angle.

We will then tackle, in order, the problems of

– the *stability in the presence of noise*,

– the *choice of the monetary base*, and finally

– the *effect of power in risk*.

Then we can analyse in more detail the discussion of speculation.

8
Illusions of Chance

Stability and Instability in the Presence of Noise

One of many questions hanging over the financial economy is whether the behaviour of speculators leads to a stabilisation or destabilisation of the markets.

The crashes of 1929 and at the end of 1987 were due to a panic whose cause sometimes appeared diminutive[1]. They unwound in a precipitous manner until a certain number of economic consequences or institutional decisions had occurred. The analogue is evident with a body in an unstable state where an insignificant event starts an accelerated motion, or with a physical system in metastable equilibrium where a small perturbation makes it fall into a more fundamental state. But in the daily unfolding of the markets, it is the operators who only base their observations of the paths of the price of the asset, the *spot*, in order to discover the tendencies to rise or fall, who are contributing to a collective behaviour to accentuate movements. These speculators, notably the *chartists* – of whom I will talk in detail shortly – use basic software tools that are at the disposal of all traders in the market rooms. The gregarious character of short-term anticipations of the operators is undeniable, and well known among the speculators themselves. If we do not see stock market crashes all the time, then it is a phenomenon that contradicts their mimetic behaviour. It can be a stationary regime, a period without economic innovation or outstanding media attention where these micro-destabilisations are slowed-down by a dif-

---

[1] See Orléan [71].

ferent category of operators for whom the stakes are much closely linked to the
economic significance of the assets. This means the operators who buy and sell
assets after making an economic study on the likely development of the activity
concerned, and take account of the predicted saturation of the demand and the
expected technological mutations, without making a single observation of the
daily development of the price. The economic observer of the situation incites
them to judge certain prices as unreasonable and they tend to reestablish the
price in brackets that are more normal *in their eyes*. An operator of the second
category behaves in the market by putting a point of view on the real dynamics
of the economy, and in order to do this, it seems that he must pay a sort of fee
to these attentive operators on the whim of the market.

The phenomena of instability are produced when the cause and effect vary
in the same way. In mechanics, for example, a solid body moving around a
horizontal axis whose centre of gravity is above this axis is in unstable equilibrium because displacing it by a small angle will make it subject to a couple
whose action will tend to increase the deviation. A large number of natural
phenomena are governed up to first approximation by a linear relation between
the state and the rate of the change of the state. This is very often the case
in chemistry when the rate of a reaction is proportional to the concentration
of the substance that it produces. Or in economics, when the profit given by
a financial participation is proportional to this participation and is reinvested
continuously. Close to a position of equilibrium one can usually assume that
the relation between the state and rate of change of the state is linear. If the
coefficient of proportionality is negative, the speed is directed towards the equilibrium point: the equilibrium is stable. If the coefficient of proportionality is
positive, the speed has the same sign as the displacement so that this increases
indefinitely until it meets an obstacle: the equilibrium is unstable. In the case of
a deterministic system, physicists have known since the beginning of the 20th
century how to treat these questions in complete generality[2]. Their methods
and concepts also apply in the random situation, but then they have a much
more delicate interpretation, and the conclusions split into those that apply to
each trajectory, and those that apply on average. These can be very different.

This makes the interpretation (economic or political) of these deceptive
graphs that are now so abundant in the press more difficult than some com-
mentators would like to think.

To take a simple example, consider first a SICAV[3], formed from a basket
of debentures which each have a yield of 4.5% and whose value on the market

[2] See A. Liapounoff, "Le problème général de la stabilité du mouvement", Annales
de la Faculté des Sciences de l'Université de Toulouse, 9, 1907.
[3] A SICAV is a type of mutual fund, unit trust or open-ended investment
trust. (Translator)

hardly varies, (is constant, let us say). Such a SICAV is then a good place to reinvest at 4.5% continuously. From a physical point of view, it is an auto-accelerated phenomenon, hence unstable, though its instability leads to profit.

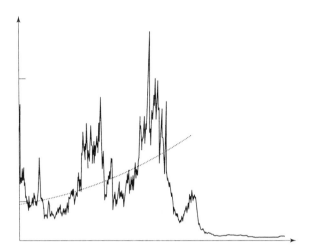

A phenomenon where the natural dynamic has an increasing and accelerating behaviour in the absence of noise can have a very different behaviour in the presence of noise. A strong noise prevents us from knowing if the dynamic is stable or unstable.

Consider now a different but analogous SICAV, formed by debentures whose market value is very variable. A noise is imposed on the preceding dynamic. The simplest hypothesis is that this noise is proportional to the value of the debentures, and one can call the coefficient of proportionality the *volatility* of these debentures as we obtain a mathematical model identical to that of Black-Scholes. If the volatility is weak, our second SICAV will behave *much the same* as the first, it will simply be more fluctuating. But if the variation is much larger and the volatility exceeds a certain threshold value, here 3%, the qualitative behaviour of our second SICAV will change. On average it has a yield of 4.5%, but this average does not appear in each trajectory. It is an abstract average that would be observed if one had a large number of independent such SICAVs. When we look at one trajectory we see that the SICAV changes enormously when the market volume is high, but which on the other hand it goes to sleep when the value is low, to such an extent that it finishes by drifting completely towards zero.

The exact values do not matter, they result from calculations using a model of noise, the details of which do not concern us. What matters is that one sees a

threshold appear beyond which there is a change of behaviour. The presence of
strong noise has mechanically *stabilised* the system, which here, unfortunately,
leads to the loss of the initial investment.

One can present this phenomenon in a more concrete way by considering an
amateur economist who attempts to curb his urge for gambling by going just
once a year to the casino. Prudently, he has taken care to invest his wealth at
10% annually and by equal prudence each year he only bets half of his wealth
on *even*. A true gambler would use a more sophisticated martingale, but we
do not want to complicate the problem. Under these conditions *he is going
to be ruined*. The cumulative *effect* of his investment and his gambling, while
equitable, or in other words zero-sum, is going to make his wealth tend to zero[4].

It is more intuitive to see this effect by looking at the case where our amateur
bets only half his wealth each year without investing his money. One will not be
surprised then of his progressive ruin if when he loses he divides his wealth by 2,
and when he wins he only multiplies by 1.5. His ruin remains the only long-term
issue if the money is invested at low interest, but the situation improves when
interest is high. The moral of this fable is then that even with an investment
at 10%, an amateur gambles too much by betting half his wealth. In the same
way, with interest rate at 4.5%, the SICAV cannot allow long term volatility
of more than 3%.

When an economic quantity tends towards zero, it leads to ruin if the quantity concerns profits, and to success if it concerns losses. In either case, noise
can give an illusion. The question is to know *from observing an evolution if
one can deduce the underlying deterministic dynamic and the amplitude of the
noise*. One can perhaps then consider that the deterministic dynamic corresponds over a certain period of time to the economic reality – the fundamental
– and find explanations for the observed noise.

Here the change of the probabilistic axes due to Girsanov gives a definitive
response. *Only the noise (the volatility) can be known precisely.* There can only
be a likelihood for the underlying deterministic dynamic, which is all the more
woolly when the noise is large.

To recap, one cannot see if a sizeable market is in equilibrium, either stable
or unstable. This property depends on the underlying dynamic and its relationship to the amplitude of the noise. To evaluate it, it is necessary to understand
the role this quantity plays in the operation of the economy. However, here
opinions frequently diverge. This is only natural. This understanding rests on
an act of faith in a set of judgements concerning the economic forces and their
likely results in the future. Many operators only see things subjectively and

---

[4] This result will still be true if his money were invested at 15% annually, but it
would be different if it was invested at 16% in which case it would grow more and
more despite oscillations.

attempt to move away from the thinking of the market. They consider that the opinion expressed by the price on the financial markets is the only real objective, and not a particular interpretation of it.

The Problem of the Monetary Base

One can represent these different beliefs, the *ideological economics* as we might say, as a priori probabilities in the future. They constitute areas of conflict among themselves. It is clear that any bank or business wants to believe that what it participates in is going to be profitable and is going to endeavour produce information that justifies its decision[5].

An essential role in this is played by currency. Even though the currencies have noisy fluctuations in relation to one another, it is different maintaining accounts in one currency instead of another. A business in yens will not follow the same strategy as one dealing in dollars. It is perhaps surprising that the best strategy is not the same for both. This is not just because of the timing of the balance, but also because certain risky situations on one side – e.g. selling dollars following the buying in dollars – will not be the same for the other and will produce different expenses for different hedges. To make these financial operations compatible under the rubric *balance sheet* does not change the fact that the optimisation of choice for the future is not the same for both.

For a bank which participates in an economic activity that is in a very localised zone of influence of a currency, the economic hazards are more often correlated with the variations of exchange rates, while the optimal projections for the uncertain future depend on the currency in which they are expressed.

A simple example will make this phenomenon clearer. Suppose that one can pay in apples or oranges. Today we have the choice between buying an umbrella or a pair of sunglasses. Their price is the same, it is one apple or one orange. If tomorrow it is fine, the sunglasses will have the price set in oranges, if it rains it is the other way round.

The exchange rate establishes itself like this:

tomorrow	if it is fine	if it rains
rate	1 orange = 2 apples	2 oranges = 1 apple

and then the prices are established thus:

---

[5] Several observers have noted that certain American businesses have dismissed personnel with an explosion of information to the press and then quietly taken them on again simply to obtain a good management report.

tomorrow	if it is fine	if it rains
sunglasses	2 apples or 1 orange	0 apple or 0 orange
umbrella	0 apple or 0 orange	1 apple or 2 oranges

Suppose that today there is an equal chance that tomorrow it will rain or be fine, so that if one reckons in oranges, the expected value of the sunglasses is 1/2, that of the umbrella 1. If one reckons in apples it is the opposite: the value of the sunglasses is 1, that of the umbrella 1/2.

If a French financial organisation works on an international scale and if its income – from businesses to which it lends or makes agreements – is mainly in dollars, it is understandable that the various subsidiaries will search for a common way of thinking and that they will be naturally attracted to thinking in dollars. This abandons French logic. But, as we have seen, the performance of an business managed in dollars cannot be converted into euros a simple multiplication. Decisions in the short term, the influence of time, the estimation of risks, all alter the probabilities. And though they can thank the Americans when the yen oscillates in a noisy way against the dollar, to think in dollars for the French and the Europeans is irrational.

It is clear, in particular, *from a strictly financial point of view*, that the single European currency is a good idea when we consider that Europe is the premier commercial power in the world. However, more than half the world exchanges are still made in dollars. Moreover finance is only one among many other important factors such as social and cultural, that must be considered when planning the construction of Europe. We will return to this in the last Part.

The Relationship of Force to Risk

Each European country taken on its own is weak vis-a-vis world capital, that moves to find the best opportunities for profit. But what truly signifies *strong* or *weak* in finance? The rules of the financial game, like those of games of chance, are the same for everyone.

It is in practicing games of chance that one understands these situations. When one can only play with a finite wealth, if a large number of bets is necessary to win, one has a strong chance of being ruined. Poker demonstrates this very well, but the analysis of this game is too complicated to serve as an example. The game that best illustrates the strong/weak problem of finance is an African game. While the market rooms are hardly concerned with this continent, a large part of which is this side of the doorstep of poverty, Africa is a source of popular mathematical games (that the universities of the ex-AEF[6] have exploited and recast in French school books after being adapted into the local culture)[7]. This is an old game from the Ivory Coast – the Gourd Game – that, to me is the best pre-professional training in the craft of finance.

The game is between two or more players. Each player uses seeds of a certain colour and all the seeds have an identical form. There are as many colours as players. At each turn, each player puts into the gourd as many seeds as he wants. The gourd is a sort of large hollowed-out melon where care has been taken to leave inside a stem on which a single seed can sit. The gourd is shaken until one of the seeds comes to rest on this stem and the player of the corresponding colour collects all the seeds in the gourd.

After this, players exchange the seeds so as to keep the same colour. There are a number of variations to complete the rules of the game such as whether each player knows the stakes of the others or not, and so on. One can play with the same number of seeds initially or not, with borrowing allowed or not, with interest or not, and so on.

One can easily see that the game is equitable; each player has an expectation of zero gain at each turn. Two strategies immediately come to mind. The first – prudent – (apparently) consists of limiting the possible losses, by playing a small number of seeds at a time, to try the long game. This is a strategy to avoid risk. The second – bold – is to gamble everything at once, a strategy of high-risk.

You can test what this means in terms of finance by noting the crucial fact that *the prudent strategy is catastrophic if the opposing player has many seeds and if he follows the bold strategy*. The prudent player soon realises, by paying, how rare a rare event is!

Whatever the chosen strategies of the players, if they start with equal amounts, then the game rapidly becomes unstable. It is clear that whoever has the most seeds has the advantage. This is not only because one player has more seeds than his adversary, but also that he has at his disposal more strategies and because he can put in more seeds if he wishes. Playing this game teaches us very quickly that *advantage lies with the player who takes more risks than his adversary*.

Here risks are taken in a particular way that is not measured by the first moment (expectation) nor the second moment (the variance) but rather by the third moment[8], and by other moments connected with it.

---

[6] The French colonies in Africa. (Translator)

[7] See Zaslavsky [87], and also S. Doumbia et al., "Mathématiques dans l'environnement socio-culturel Africain", Institut de Recherches Mathématiques d'Abidjan, 1984.

[8] See "Moment" in the Glossary, page 138.

The two tables below show the values of these moments of the law of probability of the gains of the players according to whether their bets are symmetric or not.

1. Each player stakes 4 seeds

	1st moment (expectation)	2nd moment (variance)	3rd moment
Gain on 1st round	0	16	0
Gain on 2nd round	0	16	0

2. First player stakes 2 seeds, and second player stakes 8 seeds

	1st moment (expectation)	2nd moment (variance)	3rd moment
Gain on 1st round	0	16	+96
Gain on 2nd round	0	16	−96

As we have stated, the player who bets a large number of seeds in relation to the other has the advantage, and has for his gain a very unbalanced law of probability: he loses a lot with a small probability and wins little with a probability close to 1.

As the game is a zero-sum, the centre of gravity of the law of his gains is very close to one of the extremes, i.e. it functions *like a lever*.

This phenomenon is profound and is intimately linked with the concept of risk. This has been known for a long time and has been carried to an extreme form with the *paradox of St. Petersbourg*, that we can state as follows:

> *How much are you prepared to pay to play in the following game? You toss a coin, and you receive 2^n dollars if heads is obtained for the first time on the nth toss.*

This question was raised by Nicolas Bernoulli in 1713 and studied by Daniel Bernoulli in his Memoirs of the Academy of St. Petersbourg[9]. Note that the game that you intend to play can only bring in money. You have to pay a fee to play. The question is: how much are you prepared to pay to play the game?

If you choose $1000, you put yourself in the following situation: if heads occurs before the 10th toss, you lose money ($488 at least). If heads occurs for the first time on the 10th toss or later, you win something. One can clearly see

---

[9] On the history of the St. Petersburg paradox and its connections with the foundations of decision theory, see G. Jorland, "The Saint Petersbourg Paradox", in *The Probabilistic Revolution*, eds L. Krüger, L. J. Daston, M. Heidelberger, MIT Press, 1987.

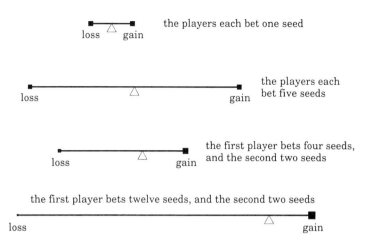

the players each bet one seed

the players each bet five seeds

the first player bets four seeds, and the second two seeds

the first player bets twelve seeds, and the second two seeds

The probability law for the gains of a player can be represented by a see-saw. In the case of a balanced game, the see-saw is in equilibrium with its pivot in the centre, as in the gourd game. If one of the players bets more seeds than the other, the see-saw is very unsymmetrical, producing a lever effect.

---

a bad configuration of risk: you win a lot with small probability and you lose with probability close to 1 (here 0.998).

It is here that probability theory appears paradoxical, since according to the criterion of the expectation of profit, you must expect to pay an infinite sum to play this game.

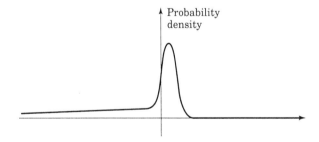

The game of St Petersbourg shows that probability laws, where with a small probability one loses a lot, are configurations of risk which are investigated much more than contrary situations.

In other words, in the sense of classical rational thinking, one gains a higher advantage if one proposes paying a million dollars to take a chance on this game!

These considerations are absolutely basic in finance.

It is clear that large levers introduce exceptionally elevated losses. They hover over the markets or over the banks with more or less devastating consequences. This is all the more worrying at a time of frequent bankruptcies, when one fears a prolonged *domino effect*, and some collapses of prices because of a transmitted panic. Several warning shots have already alerted us:

- the Mexican crisis and the collapse of the peso in December 1994 and January 1995,

- the bankruptcy of Barings in February 1995,

- the losses of Daiwa on the bonds market in September 1995,

- the losses of Sumitorno on the copper markets.

Each case has its own specifics, and merits greater study. The media, regularly reporting in a more or less alarmist fashion, see apocalypse[10]. For all that, it is convenient to emphasise that the bankers, the treasurers of businesses and the finance functionaries are all very aware of these risks, and work to put controls into place.

These controls aim to limit lever effects wherever they appear. First, there are controls at the national level, in the context of the management of the monetary mass by the central bank, where the banks are required to respect a relationship between the stocks they possess and the loans they grant to individuals and to businesses[11]. At the international level, recommendations more and more strict and more and more detailed are put into place[12] so that the important financial establishments can control their risks. It is certain that derivatives merit particular attention in the analysis of risk precisely because they allow very unsymmetrical configurations: "One of the most controversial points about derivatives is the *lever effect*: a limited amount of money allows the purchase a very important exposure to risk of the market"[13]. Without

---

[10] Cf. for example Susan George, "Faillites du système libéral, le danger d'un chaos financier généralisé", Le Monde Diplomatique, July 1995.

[11] In works on prudential control, the term *lever effect* is employed in a more general sense than that here, close to the strong dissymmetry of the laws of probability, and includes the strong dissymmetries in time between the sums involved (weak) in order to take the risks in the future (exaggerated).

[12] We cite the works of the Comité Cooke then the Comité de Bâle (Lignes directrices pour la gestion des risques liés aux instruments dérivés, July 1994; Surveillance prudentielle des activités des banques sur instruments dérivés, Dec 1994), du Comité Technique de l'OICV (July 1994).

[13] J. Scheinkman, "Produits dérivés et effet de levier", in B. Jacquillat and J.M. Lasry [54].

wishing to stop anyone speculating in his own way for his own accounts, the banking establishments who manage for others have an ethical responsibility that obliges them to offer prudent management, allowing them to maintain their image. These methods consist in the first instance of *reacting* (usually daily) to the market price (*mark to market*) of all the assets held[14], of calculating the *sensitivity* of this position to variations of the market over a reference period (a day, three months, etc.) and then finally of defining *bounds* on permitted risks. As the market is restless, *volatilities* come into the calculations that one can learn about in various ways, historically by producing statistics on the prices and their correlations, or by a direct instantaneous measure of the agitations, or even by indirect measure (implied volatility) in making use of the market price of derivatives, that gives information on the opinion of the market regarding the volatility of the underlying. The bounds assigned to the risk are most often of the form *loss greater than k with probability less than p* but can be translated into a system of more elaborate constraints according to the various bank strategies[15].

First a remark of common sense emerges. This regulation would not be necessary without a natural propensity of speculators to take risks, that reinforce the inherent advantage of the lever as described in our analysis of the Gourd Game.

Second, we observe that methods of prudential control, whatever they are, ignore very small probabilities. The very rare events pass through the net for a large number of reasons: one does not know the queues of probability laws and one adopts the Gaussian hypotheses even when they do not seem to be quite valid; the correlations are not well known and the moments of order 3 are even less well known; finally the rare events in the classification methods are not describable since the combinatorics that defines them is too ramified. But with derivatives the losses are potentially unlimited.

These two remarks explain the headlong rush of traders who take more risks than their general instructions imagine and even more risks than they imagine amongst themselves[16]. As in the Gourd Game, they have a tendency to *configure their risk* in such a way that they will have some profit with probability close to 1 but a significant loss with a very small probability. This rare event, from hedge to hedge, can be reworked and spread over several assets, several periods, and several regions of the world so that it seems to have disappeared completely. The operator then thinks that he has realised an

---

[14] What is clearly sufficiently different from this is the increase of an effective sale of all the assets (as an offer of takeover bid does not correspond to the stock-market capitalisation value of a company).

[15] Their use in the matter regarding prudential management is notably presented and discussed by D. Cherpitel, R. Litterman, P. Reyniers and S. Migrahi in [54].

[16] N.B. and B. Walliser, "Fuite en avant", Le Monde des débats, July-August 1994.

astute speculation in every case if the rare event produces a sufficiently large and exceptional perturbation of the markets so that his personal responsibility is out of question. In the case of Barings, it seems precisely that this spreading of an unfavourable case had not been the only preoccupation of Nick Leeson who managed a position on the Nikkei index between Tokyo, Osaka and Singapore. This headlong rush lies at the bottom of the perfect success of a trader, the profit arrives almost certainly and the losses vanish in impalpable events.

From the examination of these difficulties, we are ready to learn several lessons. First, the study of stability in the presence of noise has shown us that this changes the behaviour of auto-accelerated systems. It can modify their stability. The observation of a noisy dynamic easily gives us the amplitude relative to the noise, the volatility, but does not give much information about the underlying dynamic. In particular, it is an illusion to want to find the fundamental by a single mathematical manipulation such as the moving average or some other smoothing. The changes of probabilistic axes following Girsanov show that, over a fixed period, markedly different fundamentals of tendencies can be associated with a given trajectory.

Second, we have seen that the monetary base is not a simple unit of physical measure. One can pass as often as one likes between the Imperial system and the metric system of measurement without changing the laws of physics. But when the yen and the dollar fluctuate with respect to each other each with a component of random noise, the optional decisions of a business are not the same in one currency as in the other. There is in some way a *Weltanschauung*[17], that is slightly different in the two cases. Clearly that will not happen when the exchange rates are fixed, and conversely if the volatility of the exchange rate becomes large, the struggle between the two visions of the world is exacerbated and becomes more and more costly.

Third, we have analysed the effects of force on risk and seen that the risks imposed on the market gives a strategic advantage to large intervenors over small ones and obliges the latter to some protection. We must be careful not to rush too hurriedly into the political consequences of this remark, if only because certain very large interventions like pension funds are coalitions of small investors. We have seen also that the effect of force materialises with a particular configuration of risk presenting an important lever effect, which incites the traders to conceal and ignore very rare events in their headlong rush. These the prudential international controls, while timid and a little watertight, try to limit.

After this analysis, we can now turn to speculation, the principal object of this Part, and reveal that there are in fact several different activities that form

---

[17] A philosophy of life. (Translator)

a durable structure[18]. One can distinguish three areas of speculation:

− economics,

− psychology,

− mathematics.

We will start with economic speculation.

---

[18] The term *topic*, which in psychoanalysis denotes a theoretical model of psychic behaviour, is particularly well suited since, as we will see, the speculators operate in the sub-consciousness of other interveners.

Three Types of Speculation

Economic Speculation

Economic speculation is an evaluation over the medium term, based on a knowledge of the economic life, of the quality of management of a certain businesses or the vitality of a certain section of the market. It is supported by the conviction that applied economic science can, by analysing the market, be applied to real situations to furnish pertinent explanations. Perhaps it was possible in December 1994, by means of a study of the producers, of the level of stocks, of the demand of consumers and the character of the market, to anticipate that the quoted value of white sugar at $430 per tonne was going to fall in January 1995. To take such a *position* in selling futures would return just $35 for $430 invested over one month. Why did the price of coffee increase by 135% between November 1993 and November 1994, and that of soya decrease by 20% at the same time? Some specialists profess to understand. The factors affecting agriculture, the systems of commerce and transport, as well as the movements of capital from one market to another, constitute an interpretation of the economic life that the anticipations of speculators express in the prices.

Economic speculation is based on a projection over time, e.g. one month, three months or more, a period which is sufficient for one to consider that the economy might change (the economy, in this case, meaning that of goods and exchanges that lie in a mist of disorderly fluctuations of the market). Often, in the works of the classical economists, one projects this particular acceptance on a set of speculative activities[1]. One then understands why this is presented positively: if the development of the price of coffee and that of soya were totally incomprehensible with the tools and concepts of economy, then it would be pure randomness that rules, and the search for economic science merely a ghost-hunt. It is incontestable to say that one part of investment is made after economic analysis even if one does not know which part.

Large movements of capital are usually the expression of economic speculation. It is estimated that in 1991 the two hundred most important pension funds – American, European and Japanese – were controlling 8000 billion dollars compared with only 500 to 600 billion held in the official reserves of these countries. If a significant part of the capital held by pension funds is held in one currency to the detriment of another, the set of regional monetary systems is destabilised. The situation is simple and by no means paradoxical: certain intervenors have such a weight that their interventions modify the economy slightly but significantly. Their freedom of initiative gives them an advantage over others, with their anticipations, that only they know.

When capital flees a country, the currency collapses and the capital becomes more attractive – as we saw the sterling crisis in 1992 which made a celebrity of the speculator George Soros. This phenomenon, once started, is auto-amplified. Under these conditions, the decisions of the large interveners, when they are taken at a time when the political and economic life of a particular country is marked by a significant new fact, are followed by a cortege of imitators who accentuate the effect of their initiative. If a particular mutual fund leaves the market, for example after an increase of the minimum wage of the main country concerned, and moves to another market where a privatised national company is sold, the intervener sells before the prices fall and buys *before* they start to climb. One can repeat this operation indefinitely by a well-managed route and realise large profits. This scenario can take various forms that one can group under the generic name *Topaze effect*[2], to pay homage to the dynamic municipal manager invented by Marcel Pagnol, who wanted to put urinals successively in front of all the cafes of the town[3].

---

[1] Such is not the case for Keynes who proceeded with a clear conceptual separation by reserving a particular word for the analysis to long maturity date: "If I may be allowed to appropriate the term *speculation* for the activity of forecasting the psychology of the market, and the term *enterprise* for the activity of forecasting the prospective yield of assets over their whole life, it is by no means the case that speculation predominates over enterprise. As the organisation of investment markets improves, the risk of the predominance of speculation does, however, increase." Keynes [57].

[2] See M. Pagnol, *Topaze: pièce en quatre actes*, Fasquelle Editeurs, 1930. (Translator)

[3] Some economists consider that there is a theory of capital movements that picks up the case that we mention as well as others. These ideas, adaptations of analytical thinking laboured for the optimisation of a portfolio, are clearly interesting but

Psychological Speculation

Psychological speculation is a less dramatic activity and requires much more attention to detail. It is based on the fact that the traders are human, and so it must be possible to separate out their individual actions. These speculators, called *market psychologists*, hardly ever raise a cultural or psychological thought on the nature of the tendencies and psychic faculties that are at work in the activity of operators. We will consider the markets from this point of view in the last Part of this book. These speculators are interested exclusively in quantified psychological models and in the interpretations of financial dynamics opening the possibility of profits. These present themselves in the literature as a mixed rag-bag of historical considerations, famous citations such as that of Charles Dow, founder of the *Wall Street Journal* – "buy on the rumour, sell on the news"[4], of natural principles (the Fibonacci numbers occur often in nature) of models of mimicry of starting tendencies, etc.[5]. Certain collective behaviour is evident but to take advantage of this is not without difficulty as it is often dangerous not to do the same as everyone else. An example will help to illustrate the turn of events of this type of speculation. If during the two days of the weekend a certain asset is not quoted, then when the market opens on Monday the quotation is going to arouse reactions. If the first movements show an increase, many traders will think that the analysis made at the end of the previous week concluded an increase and so the movement is going to be accentuated during the following minutes, where there is a possibility of profit. It is evidently advisable to use statistics from ones own side to practice this type of speculation, since the practices and spontaneous reactions in Chicago are not exactly the same as those in Frankfurt or Tokyo.

If one compares the market to traffic, the speculators are the speeding drivers who want to take less time than others. They are very alert, leaving as soon as the lights turn green, accelerating their car rapidly up to the speed limit (or beyond), and when their actions thin the traffic, they are good drivers. Do the traffic engineers confirm this way of looking at things? The response must be qualified: although because of these speculative drivers the traffic is quantitatively a little improved, most users prefer less risks even at the expense of a lower speed than average. When the services of the Ponts et Chaussées

---

their epistemological value is problematic in a way that the managers of funds do not believe (Soros [81]), and who see in the real situation much less rationality.

[4] That is to say, to take a position when a rumour appears and to conclude when the news arrives.

[5] See for example Tvede [83]. This literature is more pragmatic: *to say that which works.* In truth one must attach more and more interest to that said by a talented practitioner. But this is no more often revealed than in private circles. We will return to this aspect in the next Part.

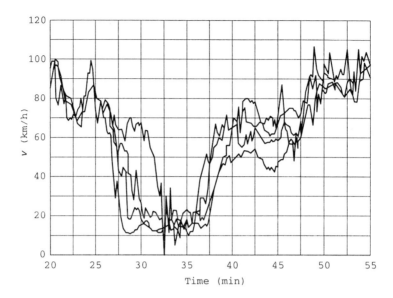

Speed measurements of vehicles as a function of time. These curves resemble those of prices on the stock-markets. If the speed varies little, or slowly, the traffic flows regularly, and safely. If the speed varies a lot, the flow is turbulent, risks a re increased, the same as when there is a strong volatility on the markets.

installed display screens on the pérépherique[6] around Paris indicating the predicted times from Port d'Orléans to Bercy or Porte Maillot to Porte de la Chapelle, etc., the behaviour of drivers changed. This experiment is very interesting for the teaching of psychology[7]. The flow became more regular. Drivers who regularly changed lane to lane in order to go as fast as possible have disappeared, or at least they have become more rare, and the displayed messages, in consequence, have gained reliability. Investigations have shown that users as a group see these arrangements very positively and most consider that they have lead to a significant improvement of traffic flow. But the measurements carried out show that this improvement cannot be judged by the total daily flow on the Paris pérépherique because that has been slightly reduced. The speculators who profit by the passive attitude of others – the elderly drivers – see improved results when the global performance is very light, but they reduce the perception of the rest of the users of the quality of traffic.

---

[6] The Paris ring-road. (Translator)

[7] See the articles of Th. Vexiau and de S. Cohen in the Annales des Ponts & Chaussées, no. 78, 1996.

Finance	Traffic
Spot value, instantaneous price	Speed at a point at a time of stream of vehicles
Volatility	Standard deviation of the speed of the stream
Transaction profit	Brevity of time of a driver's journey
Speculator	Driver who looks to minimise the journey time
Lever effect	Maximum speed attained during the journey
Crashes of price collapse	Hold-ups, jams, zero speed

This is because there are also causes of irregularities. Because of the risks taken by drivers, from time to time accidents occur that block the pérepherique for several hours. In the absence of the screens the average flow was a little more increased but the variability (the volatility) was strongly increased. The responses to surveys can be summarised by saying that car drivers as a group present an aversion to risk, which makes them clearly prefer a journey whose time is a little longer but more certain, to a time that is on average a little shorter but subject to large variations from day to day, leading to them sometimes being late for work, and missing appointments at work and with their family.

On the financial markets, we do not know the preferences of users. It seems that the general opinion would be to prefer weak volatilities, so those activities of speculators that cause risks are diminished[8]. In this analogy, the lever effect is the analogue of the speed limit: an accident is even more serious when one is going too fast. On the other hand, to my knowledge, no-one has ever imposed a psychological rule on the market to diminish volatility, yet certain soothing declarations from eminent politicians move in this direction.

Derivatives, which are indicators in the short term and are quoted in a specific market, play the role of these screens on the pérepherique. Yet, according to certain observers, their presence would have a tendency to increase the volatility of the underlying[9].

---

[8] *"It is clear that the increasing power of lever effects contribute to the existence of micro-economic risks and to the fragility of the set of financial structures, so that a particularly prudent approach is indispensable."* J. Saint-Geours in [54].

[9] *"What do the studies say on the impact of derivatives on the volatility of the underlying? From works conducted at the COB (Commission des Opérations de Bourse) have arrived at a qualified conclusion. The derivatives do not seem to influence the underlying in the medium or long term but, under certain conditions, they increase the volatility day by day."* J. Saint-Geours in [54].

Mathematical Speculation

Finally, there is a third type of speculation, *mathematical speculation*. This is neither economic nor psychological but results simply from the fact that the complexity of the financial instruments and the models used for their management create non-linear phenomena, with some effects of curvature or of a probabilistic bias, so that there are benefits on average (expectation) not compensated by an increase in risk (variance).

To give an example: the management of the Black-Scholes model of an option on an index where each underlying has its own agitation, is mathematically incompatible with the Black-Scholes model applied to the underlyings. Quite simply this is due to the fact that the indices of baskets are additive but that the exponential of a sum is not the sum of the exponentials. It is possible to take advantage of this by taking care of the fact that the weakest fluidities of the markets for the underlyings can introduce risks and transaction costs. The existence of mathematical speculation can appear surprising. This is because finance is a much more complex game than the casino. Many products are linked together, i.e. are functions, eventually noisy, of each other, and these functions are not linear – such as the maturity values of options – i.e. they do not preserve either the Gaussian character of the laws of probability, nor the martingale property of random processes.

To be more precise, I am going to describe an extremely elementary type of mathematical speculation. I do not envisage giving the reader a revelation assuring him of a fortune. I simply hope to convince him of the existence of profitable strategies with limited risk. First, we will introduce a geometric treatment of portfolios which is useful in many situations.

Geometric Representation of Portfolios

Let us take a portfolio for simplicity composed of two assets, for example yens (JY) and dollars ($). Things will be similar if there are more currencies, only the number of dimensions will be increased. If the portfolio contains x yen and y dollars, we represent it by the point (x, y).

At each instant t, the exchange rate $s^{12}(t)$, that represents the number of dollars necessary to buy a yen, is the reciprocal of $s^{21}(t)$, representing the number of yens necessary to buy a dollar. The mutual value of the two currencies can be represented by a straight line passing through 0 with slope $s^{12}(t)$, that we will call the *exchange rate line*. Note that in this representation *all the portfolios situated on the same perpendicular to the exchange rate line have the same value*. The proof is immediate: the value of the portfolio (x, y) is

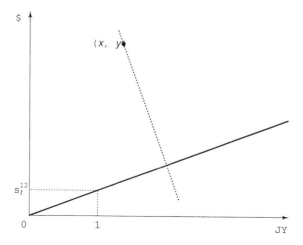

$x + s^{12}(t)y$ in yens and the line with equation $x + s^{12}(t)y = k$ is perpendicular to the exchange rate line.

Taking Advantage of the Mutual Agitation of Assets

If we start with a portfolio (x, y), then when the two assets evolve we can modify the composition of our portfolio, by selling one of them for the other.

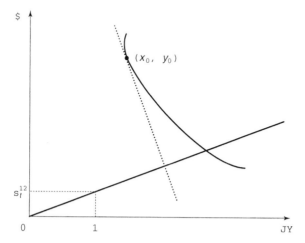

It is possible to manage a portfolio so that the representing point stays on the arc of a fixed curve and that every variation of the price gives rise to a

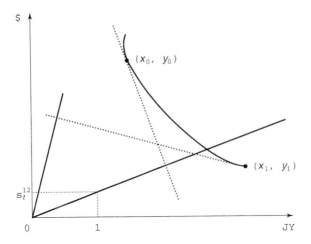

beneficial transaction.

We choose a curve that is convex towards the point 0 and such that at time t_0 its tangent at the point (x_0, y_0) is perpendicular to the exchange rate line.

At a later time, the exchange rate line has changed, then it is the tangent to the curve at another point (x_1, y_1) which is perpendicular to the exchange rate line. This tangent passes between (x_0, y_0) and the origin, that means that the transaction at time t_1 of the portfolio (x_0, y_0) to the portfolio (x_1, y_1) is beneficial.

So if one chooses a curve in the plane which turns its convexity towards the origin, and manages the portfolio by demanding that the representative point, each time it is displaced, lies on the curve at the point of contact with the normal tangent to the exchange rate line, then under these conditions *all transactions are beneficial*, and the portfolio stays permanently on the curve.

Remarks

1. If we start with the portfolio $(1/2, 1/2)$ and if we choose a curve convex towards the origin and tangent to the line through the points $A = (1, 0)$ and $B = (0, 1)$ to the point $C = (1/2, 1/2)$, this technique allows the realisation of profit by the single mutual agitation of the two assets without the deterioration of the value of the portfolio, which will always stay between the values of the portfolios $(1, 0)$ and $(0, 1)$ which are constituted entirely of dollars or of yens. Note that these two portfolios bring back nothing. This remark evidently applies similarly when the monetary units are modified.

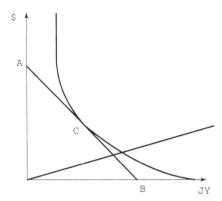

2. We can see that if the two assets are agitated but stay nearby, one has a better management of the portfolio than if each of them is taken on its own. This is the case, for example, for the deutschmark and the Swiss franc for historical reasons of stability and these currencies whether in francs or in deutschmarks take account of the wishes of convergence attached to the Treaty of the Union[10].

3. The shape of the curves and their radius of curvature can be adjusted over time. Since the transaction costs are also well-known, one can only make a transaction that is beneficial taking account of these charges.

This example of speculation is clearly known to certain teams. The statistics for this method applied to the futures market over a year show that it has benefits that without being considerable, are interesting. We add that these techniques are significantly improvable.

---

[10] In this case, this type of speculation is likely to stabilise the franc with respect to the deutschmark by diminishing the volatility of this pair of currencies.

10
Practical and Moral Values

Positions and Arbitrage

In practice the three areas of speculation are clearly not to be considered separately. They are not exclusive. There is also a fourth field – the legal field. The disparity of regulations concerning the taxation of derivatives in each country is substantial and allows a route liable to more or less tax according to the methods chosen. However, this is more about fiscal optimisation than speculation.

The thinking behind speculative transactions is mixed. Statistical analysis, that modern information methods allows us to perform very precisely, deals with information on frequencies, laws, extreme values, correlation or independence of prices. In this kind of mathematical treatment the economic and psychological causes are not considered separately. The speculators who work principally in the fields of psychology and mathematics, starting only with numerical data, are called *chartists* if their methods are graphic, and *experts in technical analysis* if they use more learned tools (spectral analysis, psychological methods of behaviour, statistics on time series, data on the volume of transactions, recognition of forms, etc.). The sophistication of their elaborate treatments sometimes surpasses the level of a third year in university.

The widespread use of the golden number and the Fibonacci numbers is significant and merits analysis.

The golden number ϕ is

$$\frac{-1 + \sqrt{5}}{2} \approx 0.6180.$$

This satisfies the equation $\phi^2 + \phi = 1$, which gives it its remarkable mathematical properties. If you divide a segment into two, such that the ratio of the small part a to the large part b is equal to that of the large part b to the whole $a + b$, then this gives the golden number:

$$\frac{a}{b} = \frac{b}{a+b} = \phi.$$

This property, pleasing to the eye, is much valued in painting and architecture, for example, that of Albert Dürer and Le Corbusier[1].

The Fibonacci numbers, discovered by Fibonacci, a mathematician in the 13th century, are defined by the relations

$$f_{n+1} = f_n + f_{n-1}$$
$$f_0 = f_1 = 1.$$

They increase rapidly and appear in a variety of combinatorial questions, and in nature (genetics, shells etc.). When n increases indefinitely, the limit of the ratio $\dfrac{f_n}{f_{n+1}}$ is the golden ratio.

Why are the golden number and the Fibonacci numbers used, and in what circumstances? Don't forget the psychology behind the actions of speculators. After a strong increase creating a significant swing all traders are convinced of the imminence of a significant low, but they ignore the amplitude of the swing. In such circumstances, predictive theories are good to follow because they reduce risks for all traders. This produces a consensus giving a certain credibility to the apparently crazy doctrines that are adopted. This is the case with the golden number and the Fibonacci numbers.

These methods work like witchcraft: they are effective because people believe in them. In the eyes of mathematicians and economists they are illusions. The fact that the MATIF in Paris, and the LIFFE in London, based on the example of certain American stock markets, functions like a noisy auction, is perhaps no stranger to these customs. The MATIF is moving, as is the DTB in Frankfurt, towards an electronic system.

In practice, the psychological and mathematical speculations used in technical analysis cannot be separated from economic speculation. If everyone were to apply these techniques exclusively regardless of the rest of the world and the economic significance of the price, the market would wander about. Without going to this extreme, however, the two activities are linked.

We come back to the traditional separation between taking a position and arbitrage. Classically, taking a position is precisely what we have called an

---

[1] See M.C. Ghyka, *Esthétique des proportions dans la nature et dans les arts*, Gallimard, 1927, *Le nombre d'or*, Gallimard, 1931; Le Corbusier, *Le Modulor et modulor 2*, ed. de l'Architecture d'Aujourd'hui, 1955.

US$/DMK — Bar 30 Min. — Mov. Av. : 24 Hrs 48 Hrs
05:37 GMT — 1.8734 — TRADER MADE 21.00

Chartism is a speculation, both mathematical and psychological, that leads to the study of statistics, graphics and frequencies of a price in order to judge its response to the market in certain situations where certain facts have been obtained. On this copy of a screen display we see the price (the very irregular curve), a moving average by day and a moving average over two days (gentler curves), so the bundles of lines are supposed to represent the tendencies.

economic speculation, it is made in the duration and it takes a risk. On the contrary, we have already employed the term *absence of arbitrage* in the sense of *absence of opportunity for profit without risk.* This is the first meaning of arbitrage: to profit from a difference in price between two products that are in fact equivalent. In practice it is not so simple, because most often this equivalence between the two products uses certain information, be it economic or mathematical, so that arbitrage is generally a risky operation similar to taking a position.

We take as an example the textbook case of implied volatility. As we have already seen, the value of an option according to the Black-Scholes model makes use of the coefficient of agitation of the underlying asset (its volatility). Because of the existence of *derivative markets*, the option is itself the object of a quotation. This quote for the underlying corresponds to a fictional volatility that one calls *implied volatility.* One can compare this to the real volatility and deduce that the option is under-quoted or over-quoted on the derivatives market. This will be to trust the Black-Scholes model more than the advice of operators on the options market. The example is simplistic but it lets us see the subtleties of the concept of arbitrage, that uses all the knowledge at one's disposal, be it economic, historical (statistics), mathematics (with models more detailed than others), and so on.

In addition, one knows from these examples that the speculator uses more and more sophisticated scientific tools. Moreover it is clear that anticipations based on economic analysis are those that are also strongly dependent on mathematics. Economic analysis is based, in effect, on models whose parameters are identified by the given treatment of econometrics. These are explicit models that interpret the operation of exchanges of services and of taking a position, or of estimative models that only use time series. In a general fashion, by delving deeper into one of the three fields – economics, psychology, mathematics – *the speculator tries to find a logic governing the financial phenomena that is outside the conscious analysis of other interveners on the market*, because only this type of knowledge is likely to make a profit. For that, he must on one hand know the usual interpretations, the common ideas that constitute the major understanding of the operators, and on the other hand to try to imagine other perspectives which will be more suitable to him. These can be subjective judgements on what one can call *the geopolitics of financial passions* (the importance of the next announcement of the unemployment figures in the United States and the reactions of the operators to the anticipation of the announcement of these numbers, etc.). But for this distancing, scientific and numeric tools are more and more a great help, for they allow us to create some routines and to find new solutions. Thus speculation has become a true outlet for science – we will return to this – and a professional outlet for scientific studies.

Now that we have a more precise idea of the diversity of approaches and of objectives pursued, we can approach at least some of the moral and ethical questions that these practices arouse. They are posed at several levels and are connected in general with power relations.

Ethical Questions

Consider the world level. Financial logic deploys itself in the international scene, imposing considerable constraints that many countries cannot support. We approach this subject from the direction of the transfer of power between the markets and the states. However, before looking at questions of political economics, an ethical problem poses itself at the international level regarding major economic facts that sometimes result from decisions taken with reference to other concerns of a speculative nature.

Too often ethical problems are overshadowed by some global financial crisis (i.e. a major crash). The systemic risk, I must emphasise here, is a very real preoccupation of the financial world. There is a great deal of important literature on this subject. The probability of an earthquake decreases strongly with its size. This must not hide the ethical problem of speculation at the world level that concerns the future of many countries. I wonder if it is possible to work so that the historic developments are not just lucky consequences, nor flaws of some sort, of the operation of the world markets. One of the reasons for which the progress of financial regulation is so slow and difficult is that one cannot think of finance as acting above the true economy. We will see in the next Part how the mechanisms of financial markets wipe out the distinction between speculation and true economic values.

They contribute in what is clearly set in a post-Keynesian perspective. In 1935 the author of the "General Theory" wrote[2]:

Even outside the field of finance, Americans are apt to be unduly interested in discovering what average opinion believes average opinion to be; and this national weakness finds its nemesis in the stock market ....

... When the capital development of a country becomes the by-product of the activities of a casino, the job is likely to be ill-done.

The measure of success attained by Wall Street, regarded as an institution of which the proper social purpose is to direct new investment into the most profitable channels in terms of future yield, cannot be claimed as one of the outstanding triumphs of laissez-faire capitalism —which is not surprising, if I am right in thinking that the best brains of Wall Street have in fact been directed towards a different object.

In other words, Keynes, at the end of his treatise on the mechanisms of saving, investment and employment, denounces speculation in the name of a healthy economy. Today the argument no longer holds so rigourously, and when he adds[3] that:

The introduction of a substantial government transfer tax on all transactions might prove the most serviceable reform available, with a view to mitigating the predominance of speculation over enterprise in the United States.

he showed a confidence in interventionism that does not have the same basis today[4].

---

[2] See Keynes [57], p.159 (Translator)
[3] See Keynes [57], p.160 (Translator)
[4] Independent of the fact that the idea of tax on transactions does not fit in with

It is very interesting to note what George Soros says on the matter:

I play the straight game. If one manages to open a breach in the rules, the fault lies with those who have made them. My position is clear, it defends itself. That someone treats me as a speculator does not lose me any sleep. It is not about defending the speculators: I have more important battles to fight. To replace a system that does not profit speculators is the role of the authorities. When the speculators make money, it is because the political leaders have failed.[5]

Soros's point of view is not so different from that of Keynes, he simply does not support an economic way of thinking[6].

The ethical questions also spread out at a second level, that of the financial management of the treasuries of banking houses or large companies. Most of the operations on the markets start with economic information. Admittedly information is more or less pertinent for the prediction of the future, but it is above all more or less accessible. We approach here the offence of insider trading. This second level raises the question of knowing whether the asymmetry of information between the interveners is likely to unsettle the operation of the markets[7].

More generally, for the preparation of decisions, and the access to data, the relations between the agents are areas in that a clarification of values is needed[8].

The moral conscience of the individual operators constitutes a third level. It interests the general public because it easily evokes a romanticised view of the banking world. This should be approached carefully while endeavouring to put aside the ideologies that cloak their subject.

---

the management of derivatives in delta neutral. As we explained in the previous chapters, this method, universally employed, depends on the many transactions that constitute the hedging portfolio.

[5] Soros [81]

[6] We cite several references touching on the financial ethics at the world level: Saint-Geours [78]; B. Larre, "L'économie mexicaine depuis 1982", Observateur de l'OCDE, Oct-Nov 1992; S. de Brunhoff, op. cit. F. Chesnais, *La mondialisation du capital*, Syros, 1994; P. Dembinski and A. Schoenenberger, *Marchés financiers une vocation trahie?*, FPH, September 1993.; P. Veltz [85]; F. Chesnais (ed., *La mondialisation financière*, Syros, 1996; P.-N. Giraud, *L'inégalité du monde*, Gallimard, 1996; also the magazine "Finance et développement", published three times a year by IMF and the World Bank, that tackles regularly this thematic notably the issue of December 1995 on financial markets.

[7] For a discussion of the role of insider traders on volatility, see Artus [7].

[8] See Hélène Ploix, *Pourquoi et comment, aux Etats-Unis, se met en place progressivement une formation à "l'éthique Financière?"*, Revue d'Economie financière, no. 22, 1995.

It is from the economic, psychological or mathematical angle – and we have seen that speculation intervenes in these three fields at the same time – that the particular talent of speculators poses a moral problem. To undertake this it is helpful to make an analogy with lawyers. In a right-wing state, one accepts that good lawyers are mobilised even for bad cases. There is a balance of justice. Yet the *talent* of a lawyer is problematic in itself, since although in the end it allows him to win, it does not matter what he wins or for who. The inequality of talents is likely to bring a bias to justice. The legal profession is one where ethical thought is old but in permanent development.

What then is on the side of speculators? As a mathematician, and because I have had the occasion to organise contracts on the probabilistic analysis of derivatives, I have had access from time to time to the market rooms and to specialist agencies. There one meets some very intelligent people, from various backgrounds, many of whom were taught at schools for engineers and business, but who have often had no other experience of life in industry or in commerce, even less in politics or societies at the local level, and who spend most of their time in front of computer screens. My observation is that the most skilful make enormous amounts of money for their organisations by their *astuteness*[9]. Who pays? The model of the metro tickets[10], although very basic, is in fact correct: those who make the same type of operations but for different reasons. The speculators are also victims of the lucky finds of others unknown to them.

In these situations where we have met some moral problems and where the levels by which they arise have been roughly worked out, is it possible to clarify the statements of the ethical choices to be resolved?

Here is an old philosophical problem: the formulation of ethical questions sends us back to a grid of values. This grid contains the problematical cases that one often meets in appealing to the definition of ethics in the various professions, as well as the exceptional cases relating to religious principles, systems of political thought and ethical philosophy.

So it is important to stick with human actions in relation to the financial markets and to the difficulties that their actions create both for themselves and for others. The most concrete questions seem to me to be the following:

---

[9] And also in making money themselves. The results of a bank are very sensitive to the quality of work of the traders as is confirmed by their use of the bonus; *"the salary of a senior trader (more than five years experience) spreads between roughly 350,000 and 650,000 francs annually. To this a bonus is added, calculated from various formulae linked to personnel objectives proportional to the results of the bank, from the trading room or from the office desk. Ceilinged in some establishments (at 50% of the fixed salary for example) others are unlimited and even allow the best to gain several million francs ... so as not to be tempted by the sirens of London or New York."* (A. Chaigneau, "Les activités de marché à l'heure de l'aggiornamento", Banque, Jan 1996.)

[10] See page 48. (Translator)

1. Is money an indifferent fluid without memory?

2. Should one know the author of a transaction?

3. Who must pay for the risks due to the mimicry of behaviour?

The first two questions are linked. The first depends on knowing whether a financial operation emanating from an earlier fraud is legitimate or not. Is it the same in every case? In what measure are the consequences the responsibility of the financial players who have been deceived? The case of drug money, that on a world scale is considerable, is an important example of this.

The second is to know if the practice of anonymity, which is common because of the use of intermediaries, is not harmful to competition, and even whether the precise relationship between the author of the transaction, his shareholders, and their alliances should not belong in the public domain. That would help the fight against wrongs and allow the pursuit of audacious agents in the same way that the insurance companies apply the system of no–claims bonuses against imprudent drivers.

The gregariousness of behaviour is possible because the origins of the transactions are secret. In industry, an idea leading to new processes can be protected by a patent. Although this is not sufficient to prevent forgery, it does allow its suppression. On the contrary, as we have already seen, an arbitrage is, by its nature, an idea which disappears if it is revealed. Thus, the markets are deprived vis–a–vis of cheats who imitate a reputable operator for his methods of analysis. For the moment the risks – which are either weak and frequent or strong and rare – seem to remain in the charge of the slowest operator as in the game of musical chairs. Have derivatives the tendency to calm the game or weaken it? Opinion is divided.

The search for clear ethical principles on which to construct international controls is made all the more necessary as the current confusion has many consequences on the political and cultural decisions of the utmost importance, amongst which I cite:

— the recent influence of financial speculation on the ratio of the remuneration of capital to that of work;

— the legitimacy of the shift of political power towards the financial markets caused by the debts of states and local communities;

— the marketing of information and the question of an economy of goods and of exchange of private knowledge.

For economists and mathematicians this last point touches scientific ethics in a similar way to that for research workers in biology.

In the next chapters, we will continue the discussion of these three points. We will examine whether economic theory can provide judgements that are objective and universal on good allocation of financial resources, or alternatively if finance must be considered an area where knowledge is naturally private. We will then go deeper into the operation of financial markets, to release, by some comparisons, the profound nature of power at work in the current world financial order.

Part IV

The Stakes and the Payoffs

11
Markets and Economy

In this chapter we shall revisit several ideas that have been raised during our journey into the mathematics of risk and finance, without getting too involved in the technical aspects.

Regarding the casino, what philosophy do we retain? Secret martingales fascinate the gamblers. They are often based on false ideas, the principal one, which is widespread, being that chance obeys a certain equilibrium over time. This often creates the atmosphere in gaming rooms[1]. Sometimes they are based on reasonable assumptions, but often the apparently correct martingales have some hidden defect that is revealed by probability theory. This allows us to state in every case that it is that the strategies that allow us to win, of course, but in a number of cases that depend on chance and which are not a priori limited, such that when he thinks that victory is assured the gambler always ignores a small number of possible cases, which are precisely the cases where he will be ruined before being able to complete!

If the number of parties is fixed in advance, then whatever the strategies of the gambler, the expectation of winning is zero. This is what characterises a pure casino. The mathematics of a casino have reduced martingales to the rank of fantastic illusions and have given birth to a fertile theory, which studies the random processes that represent the cumulative profits of the gamblers in pure games. They are characterised by the centre of gravity property: the current value of the process is the centre of gravity of the values at a later time, taking

---

[1] For example, a player who consciously notices each bounce and declares loudly that the appearance of zero had only happened once yesterday.

into account their probabilities.

This property has rich consequences and, with a certain irony, mathematicians have named these processes *martingales*. This is legitimate in some sense, since the other martingales, the dreams, have no right to be mentioned in the context of science. So, born from a generalisation of the idea of Brownian motion, the theory of martingales developed after the Second World War through to today as an accumulation of results, giving rise to a new theory of integration, that found a spectacular outlet in finance during the 1970s. This brought to games and to the markets, which are a sort of game, a clarification and a generalisation of the concepts that allowed a large development of models: good for the hedging of options and the management of a portfolio, and the theory of games and finance economics.

Grounds for Belief and the Theory of Utility

This description leaves aside, however, the question of the *grounds for belief*. At a pure casino the expectation of gain is zero. If you play a game, is the way you play immaterial? The answer is no, as the discussion of the Gourd Game clearly showed. If at the casino one player puts $3500 on number 36 and another player puts $100 on each number except 36, both have a possible loss of at most $3500 and expectation of gain 0. But the distribution of their risk is very different. The first player wins a lot with a small probability, the second wins a little with probability close to 1. The dissymmetry is favourable to the second player, unfavourable to the first.

It was in the 18th century, following the writings of Gabriel Cramer (1704-1752) and Daniel Bernoulli (1700-1782) on utility, and the discussions of points of view of Buffon following Diderot, and Laplace, and Bayes following Condorcet, that the concept of the probability of an event was seen as a legitimate and important topic for philosophical investigation, i.e. its interpretation in terms of the grounds for belief that it will occur. This question lead Condorcet and Buffon to note that one forms an opinion without taking account of events with small probability:

> So the grounds for belief that if over ten million white balls are mixed
> together with one black ball, then it will not be the black that I will pick
> first, is the same kind of grounds for belief that the sun will not fail to
> rise tomorrow ...[2]

[2] J.A.C. de Condorcet, Essai sur l'application de l'analyse à la probabilité des décisions rendues à la pluralité des voix, 1785.

This problem is not dissipated by mathematics and is worth being examined more deeply since certain risky behaviour of traders depends on it. An undecided question for example is the following. If, in our casino, the second gambler wins, (which is very likely since he has 35 chances out of 36) does he think of playing in the same way in a new game, or does his taste for risk become blunted? It is clear that if his strategy consists of playing until he loses, his configuration is no longer favourable to him.

As mathematical probability theory leaves open questions about the grounds for belief, economic theory has taken over to try to take account of a certain rationality in the behaviour of the participants who must make their decisions in situations where uncertainty reigns. It is about the modelling of psychology, clearly very basic, by a rigorous formalism. Following the ideas of Daniel Bernoulli and Cramer, or authors such as T. Barrois[3], F. Ramsey[4] and above all J. Von Neumann and O. Morgenstern,[5] a theory of utility has been elaborated that tries to take account of the preferences of the economic agents amongst the baskets or even, in a random situation, amongst the lottery of the baskets. Under the hypothesis that a certain logic is followed by the agents, there exist utility functions which balance the choices in a way that the agents take the decisions that maximise their expectation of utility. To find the strategies of agents is an optimisation problem. Numerous mathematical techniques are then available and one can obtain solutions to a variety of problems.

Evidently criticisms and improvements were brought to the theory of utility. The axioms of this way of thinking are contestable. First, they do not apply to decisions taken at meetings where the choices, as Condorcet has shown, do not have the property of transitivity. Many other paradoxes have been discovered[6] leading to a framework of thought known as the *Theory of Games*[7].

These utility functions, that are supposed to govern the propensity or aversion to risks of the agents and allow delicate calculations of optimal strategies, are always poorly known both by the modeler and the agent himself. Moreover they are susceptible to the general criticism that the attitude of agents vis-a-vis rare phenomena – which are very significant – is the object of a specific line of thought taking account of the significance of the event in question, the circumstances and the peculiarities of the situation and the complexity of the

[3] T. Barrois, "Essai sur l'application du calcul des probabilités aux assurances contre l'incendie", Lille, 1884.

[4] F. Ramsey, "Truth and probability" in *The foundations of mathematics and other logical essays*, London, 1930.

[5] O. Morgenstern and J. Von Neumann, *Theory of games and economic behavior*, 1944.

[6] Cf. notably Walliser [86] and "les paradoxes de la décision rationnelle", Annales des Ponts & Chaussées, no. 76, 1995.

[7] Cf. on this subject, A. d'Autume, "Théorie des jeux et marchés", in Formes et sciences du marché, cahier d'économie politique, L'Harmattan, 1992.

relations between the agent and the surrounding world. In other words, the subjectivity of the grounds for belief goes well beyond the freedom to choose a utility function – it *resides in the interpretation of events on which the action is based.*

For example, assuming the accuracy of the facts and numbers we read in the press, to have lost six billion francs at Barings, Nick Leeson was condemned to six years in prison. But he had concealed more than a hundred billion francs in accounts in Germany (Le Monde 2/7/96). Clearly the unfavourable case was not evaluated in the same way by Leeson and his employer. Risk comes from an important problem for society: what risks are taken, who decides and in what configuration? The accepted risk is the weakest form but also the most fundamental social link. The statistics on road accidents and on illnesses disturbs and troubles us, not only because these reinforce our fear of these events but, above all, because if something happens to us we are uneasy: was it our fault or that of the society we live in? We willingly expose ourselves to certain risks, such as driving, but we do not like to expose ourselves to risks we do not understand[8]. This attitude to risk is obviously an aspect of problems of the environment.

Let us return to our casino, and compare it with horse racing, that presents an attraction of remarkable permanence. On one hand, we have a pure game (without tax) where the mathematics can be applied, on the other a game in which the utilisation of information on the stud farms, the jockeys, the nature of the horses, the ground and so on, allows us to improve our chances and to appear smarter than others. For the casino, the lottery and so on there is no regular press, as there is not a lot to say. Papers abound on horse racing, however, and in this regard the stock markets are analogous to the races. One does not know if there are biases up or down, but the specialist press gives us figures and information. With the money-traders, just as with the race-goers, such a magazine grants us the confidence of a so-called expert. But we remain confused. The financial markets are much more entertaining and passionate. There one plays in a large arena, where there is economic analysis, mathematical and psychological analysis and the interaction of the professional expertise.

All resides finally in the fact that one ignores the tendencies of a price on the stock market. As in the case of the pure casino, where an external observer cannot decide from a finite number of plays if the gains and losses are equally likely or not, the observations of a price over a certain period allows us to know

---

[8] It is significant that the idea of drawing lots proposed by the Conseil National du Sida (AIDS) for the distribution of Ritonavir where the quantities would be insufficient, was violently rejected (Le Monde of 1/3/96) because it meant drawing lots for the ill who would not benefit.

its agitation (volatility) but not its tendencies. As we have already remarked with respect to stability in the presence of noise, this is because of changes in the probabilistic point of view. The same reality of the market can be described independent of the various axes. It is a principle of relativity.

Change of Axes in Physics: The Experiments on Light

The importance of this principle to financial economics is comparable to that in physics. It may be interesting to skim through this chapter on physics to see how the plurality of points of view is important in science and in the history of knowledge.

After Hook's experiment on the speed of light in a medium (1868), it seems that Mascart, Professor at the College de France, was the first to affirm that all attempts to detect a movement with respect to the ether would lead to failure. On the basis of a series of experiments of remarkable precision[9], that would convince later theorists, he concluded that

> ... the movement of the rotation of the earth has hardly any appreciable influence on the optical phenomena produced with a terrestrial source or with sunlight, so that these phenomena do not give us the means to appreciate the absolute movement of a body and it is only the relative movements that we can obtain[10].

After the negative result of the experiments of Michelson and Morley, that confirmed the ideas of Mascart, the physical *principle of relativity* was proposed. It only became a coherent theory after the formal and conceptual work of Lorentz, Poincaré and Einstein starting from the equations of electromagnetism established by Maxwell. It was at the famous conference of Poincaré at Saint-Louis (US) in September 1904 that the principle of relativity was named and announced as follows:

> *The laws of physical phenomena must be the same for a fixed observer, as for an observer moving uniformly, so that we do not have and cannot have*

---

[9] Described in the two Mémoires des Annales de l'Ecole Normale, 2 ème série, pp. 157–214, 1872 and pp. 363–420, 1874.

[10] These words are a translation of those of Mascart (1874). In other words, "at this time, Mascart suggested that a light ray as a dynamic, it was impossible by means of an experiment that, and in which order it is, to distinguish a privileged Galilean system."(M.A. Tonnelat, Histoire du Principe de Relativité, Flammarion, 1971).

any means to discern if we are or are not making such a movement[11].

The invariance of the laws of physics under change of Galilean axes amounts to saying that physics is conserved by a *group of transformations* that, in this case, is not the group of translations, since the speed of light is constant, but the Lorentz group. Discovered by crystallographers in connection with the symmetries of isomers, the group of transformations that leave physics invariant is fundamentally important in quantum mechanics and is even a tool for research[12].

In finance, the possibility of exact hedging is the key. It is the crucial fact which raises a problem and, as in physics, exposes the inconsistency of the experiments in relation to the ether.

Change of Axes in Finance: The Experiments on Brownian Motion

Consider a call of three months on the dollar. If I predict that an increase is likely and you a decrease, we nevertheless manage this call in an identical way, in realising the only portfolio that permits exact hedging. The argument of non-arbitrage proves that the correct hedge of a derivative does not depend on the probabilistic reference point chosen. There is then a physical financial reality that is the same under all changes of probabilistic reference points (that form a group analogous to that of translations). *This reality is the market.* More precisely, from a certain reference point the price is increasing, from another it is decreasing, while from a third point of view it is a martingale – in other words from the third point of view the market will be efficient. The efficiency is thus a property that depends on the point of view, not a property of the physical reality of the market[13].

On this question, it is necessary to look again at the analogy between prices on the stock market and Brownian motion proposed by Bachelier. This idea (several years before the work of Einstein and of Smoluchovski and the experiments of Perrin) was illuminating and has been largely confirmed by practice and current theories.

[11] H. Poincaré, Bulletin des Sciences Mathématiques, 1904, p. 302.
[12] See E. Segre *Les physiciens et leurs découvertes*, Fayard, 1984; Holton [49]; M. D. Davis, *Game Theory*, Basic Books, 1970; Segre [80].
[13] We will examine a little later the concept of efficiency and of its various meanings in detail and we will see that there is no privileged super-probability (that would play the role of the ether) with respect to what one should think of as random phenomena in finance.

However, two criticisms were made of Bachelier. The first was that he adopted the Gaussian hypothesis as if it were the only one possible. We have seen that this is not the case, indeed far from it. The second was that it was a way of expressing mathematically the fact that "the market believes, at a given time, neither in the increase nor the decrease of the true price"[14]. In current terms, for this assertion to be satisfied, he takes a centred Brownian motion as his model, that allows him to continue by saying that the mathematical expectation of the speculator is zero.

This is not quite what is needed to convey the idea. In the light of the argument of non-arbitrage and the principle of hedging, we are limited today to a more subtle treatment by the fact that the market, at a given time, believes neither in an increase nor a decrease. The price is like a particle in a fluid subject to thermal shocks, animated by a Brownian motion, but where one knows neither the value nor the direction of the gravity field. A good model belongs to some sort of equivalence class of all Brownian motions, at the same temperature.

One remembers the experiments described so clearly in the books of Jean Perrin[15] and by those which Leon Brillouin then interpreted to verify that the agitation of a grain of colloid in a liquid was due to molecular shocks, and the steps taken to confirm the theories of Einstein and Smoluchovski. These experiments were meticulous, and often needed several months of preparation using a fractional centrifuge to obtain colloids where the grains were all the same size. They consisted of studying the respective influence of the movements of grains of rubber, on one hand of the agitation due to the shocks with molecules of the liquid, and on the other the effect of gravity.

With this intention, Perrin did not observe just one particle because in addition to technical difficulties that this would have raised in order to register its movement at the time, the influence of gravity is impossible to capture without a very prolonged observation time for a slow particle with small diameter such as he wished to use. The particle has impulsions in all directions, it does not stay down when it collides with the bottom of the container, but sets off again towards the top and passes successively through all the elements of the liquid. The influence of the gravity field only makes itself felt by the fact that the particle is, on average, slightly more often at the bottom than at the top. To obtain this difference of frequency of time it stayed at the top or at the bottom, which is very weak, one must observe the particle for a very long time.

---

[14] Louis Bachelier, "Théorie de la spéculation", Annales Scientifiques de l'Ecole Normale Supérieure troisième série, t17, 1900, p. 32. The concepts of true price introduced by Bachelier is technical and is linked to the operation of the market of monthly control and has no relevance to our discussion.

[15] J. Perrin, *Les atomes*, Presses Universitaires de France, 1913; *Les éléments de la physique*, Albin Michel, 1929.

The following figure is a Brownian path such as those that Perrin observed and that he could have photographed. It is impossible to decide the direction of the gravity field from this figure. If one were mistaken about the printed orientation of the image, which cannot be detected. This is the reason why

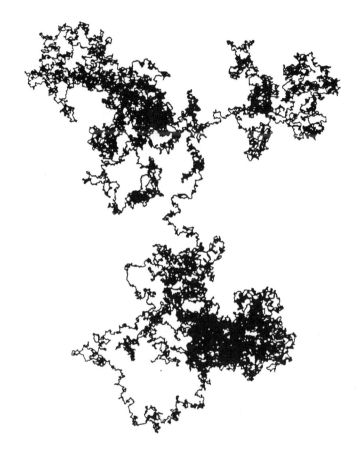

Path of a Brownian particle in the plane. It is impossible from this information to know whether the particle is affected by a field of force.

Perrin and Brillouin used *many* particles which were carefully chosen to be as identical as possible. These particles, by the independence of their movement, instantly assumed a typical state that, with a single granule, would have needed the recording of the path over a very long time. They then observed a gradient of concentration of the colloidal solution along a vertical line that allowed precise measurement, and the deduction, by the application of the kinetic theory of gases, of the Avogadro number that counts the number of molecules in a gas at a given volume and pressure.

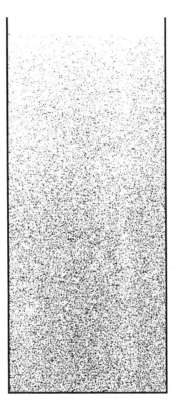

By observing a large number of identical colloidal grains Jean Perrin could detect the influence of gravity. On the contrary, if the solution contained only one particle, his observations did not allow one to know the direction of gravity unless observed for an extremely long time.

The big difference from finance is that we can only observe a single example of each stock market price, so that one cannot say if it is diverted towards the bottom or the top.

So, to continue with the Brownian analogy of Bachelier we may say that random processes are subject to a *principle of relativity* of the sort that the particles representing the price are agitated as in a liquid at a certain temperature, but without knowing anything about gravity. For an observer (speculator) who adopts a certain probabilistic reference point, this gravity is at such a value in such a direction, for another observer whose grounds for belief follow another law of probability, the force of gravity is different. For either one, however, the temperature is identical and the method of hedging a portfolio of non-arbitrage is the same.

I believe that the reader is interested in speculation and that he now agrees with me that it is subtle and complex, as are the moral judgements that are created by the techniques used. However, I cannot ignore the simplifications, even at the pedagogical end, that pass silently over some important considerations. I must also play down slightly the principle of relativity that governs finance. In physics, as was affirmed by Mascart, Poincaré and Einstein, it is not possible, through the many physical experiments carried out, to distinguish a particular Galilean frame of reference, so one cannot know whether an observer is stationary or in uniform motion. In finance, one cannot know, from all the transformations of Girsanov, which probabilistic reference is good, but one can have a very vague knowledge of it. From the observation of a price one can detect a weak probability that one calls likelihood and that furnishes ... an indication. Our analysis of the *stability* of random processes has made us understand that this likelihood is even more vague when the volatility (thermal agitation) is large[16].

So precisely, it is in return for a sufficiently significant agitation of the markets that the principle of hedging has for a consequence a relativity of points of view in finance that precludes modelling in terms a balanced game, be it favourable or unfavourable. This amounts to saying that the *fundamental*, i.e. the value of the asset corrected by the factors that would not be properly economic (modes, passions, contagion, etc.), is a concept *relative* to each point of view. This concept will be clarified when we deal with the question of the *efficiency* of markets. This has given rise to a vast literature on the subject, that can be considered in a new light.

The Efficiency of Markets is a Subjective Concept

The idea that the price of an asset on a correctly operating market is a good price, so that the agents can calculate the economic value of an asset, is a very natural idea that was formulated well before the appearance of the term

---

[16] Note that financial assets each have own their volatility. One can however classify a financial institution of an emerging market by its averages. Is there a sort of thermo-dynamic equilibrium of a market? The volatilities exchange themselves like risks through the intermediary of derivatives. There is research on this subject, that can lead to mathematical arbitrage.

efficiency. In 1889, George Gibson wrote

> *When the actions are in an open market and publicly known, the value that they acquire there can be considered as the judgement of those putting their best intelligence to work*[17].

For over a century, much has been written by financiers, economists and mathematicians about this. They first devise mathematical models of markets with excellent properties, then propose reasons to explain the appearance of the markets starting from certain economic hypotheses, and finally check the models and hypotheses using their observations of real markets in order to find the errors in their models. The first important contribution in this direction was that of Louis Bachelier, whose thesis as emphasised by Robert Merton:

> *... marks the simultaneous births of both the continuous-time mathematics of stochastic processes and the continuous-time economics of option and derivative-security pricing.*[18].

Bachelier proposed the modelling of the price of an asset using centred Brownian motion. However, his work remained ignored until the 1960s, when economists turned to the problem of mathematical speculation of the markets, notably with the work of Cootner on the risks of prices, of Markowitz on the optimisation of portfolios, and of Sharpe on pricing in the presence of risk lead to the CAPM (Capital Asset Pricing Model)[19]. This is both simple and elegant and for that is often quoted, but the influence of risk or prudence is expressed only by the expectation and variance. Therefore, a particular case of analysis generally allows uncertain selection such as Von Neumann-Morgenstern considered when the prices are Gaussian random variables or when the preferences are quadratic. From this came numerous criticisms and the appearance of new but less simple models. None of the literature in the 1960s takes account of the epistemological rupture of the principle of exact hedging and the models of current prices are of *mathematical martingales*, i.e. the expression of stable games that makes it impossible to realise a profit greater than that the market[20].

---

[17] G.R. Gibson, *The stock exchanges of London, Paris and New-York*, New-York, G. P. Putnam's sons, 1889.

[18] Merton [65]. Robert Merton received the Nobel Prize for Economics in 1997 at the same time as F. Black and M. Scholes.

[19] P. Cootner, *The random character of stock market prices*, MIT press, 1964, W.F. Sharpe, "Capital asset prices: a theory of market equilibrium under condition of risk", Journal of Finance, 19, 1964.

[20] The large part of the works of the economics school on portfolio selection reason with the aid of a law of probability that is unique for the forecasting of each agent that the modeler considers as objective. This amounts to saying that the agents have the same vision of the future and differ only in their utility functions. I

This idea was made explicit by E.F. Fama in 1965 in his *Hypothesis of Efficient Markets*: markets are efficient when the prices of assets are martingales. As the concept of mathematical martingales is – by definition – dependent on the information used for the centre of gravity property, the hypotheses of the efficient market splits into three hypotheses of increasing severity.

Weak efficiency means that the current price incorporates all the information on the previous prices of the asset concerned. If it is satisfied, a statistical analysis on the time series of the price cannot lead to any profit.

In *semi-strong efficiency*, the price contains all the information that is publicly available. The development of other assets cannot, by correlation with the asset studied, furnish possibilities of gain.

In the case of *strong efficiency*, the price contains all the economic information whether public or private. Under such a hypothesis, nobody can beat the average of the market.

As Shiller underlined in an interesting monograph[21], there is no universally accepted definition of the concept of a model of an efficient market. Nevertheless, if one were to agree on the definition of such a model, it then becomes possible to study whether the real markets are efficient or not, by means of quantitative measurements. A number of works deal with this and try to highlight the anomalies, the variety of behaviour, the speculative bubbles and so on, by the study of covariance, regression tests and other techniques[22], certain that such studies can tackle the important and preoccupying case of crashes. From the point of view of economic relevance, the concept of efficiency can be made precise and one can distinguish[23] notably an *efficiency in the diversification of risk* which relates to the complete character of the markets[24] and the important concept of *allocated efficiency* relative to the optimal utilisation of capital in its various forms. This has a less abstract meaning where it has a more operational interpretation.

The question of the efficiency of the markets then appears as a succession of attempts to model the ideal markets – ideal by their economic pertinence or because they prevent beneficial strategies. Then, by a natural scientific approach similar to one that leads the physicist to study perfect gases before real gases, elastic solids before real solids, and so on, one endeavours to see if

---

believe on the contrary that a financial market is made of fundamentally different options. This school is thus situated upstream of the rupture which means that the contingent assets can be exactly hedged and that the tendencies are subjective.

[21] Schiller [79].

[22] One of the first studies of inefficiency was by Roll and looks at the price of futures on orange juice vis-a-vis meteorological predictions, (1984).

[23] Cf. Aglietta [3], where the concepts of efficiency are clearly presented and discussed.

[24] In a complete market, all derivatives possible and imaginable can be covered according to the principle of non-arbitrage by a well-chosen portfolio.

the ideal markets modelled in this way are deducible from the hypotheses of economic equilibrium or rational behaviour. Finally, criticisms of these models have appeared through the examination of anomalies in the observed prices or the allocation of resources and more elaborate models have been proposed.

The principle of hedging brings an original perspective to this complex question. It is not an answer to the efficient/inefficient dilemma. It simply says that everyone is perfectly right to consider that certain resources are badly allocated, and that it could be more profitable to put capital elsewhere for which it would be better remunerated. It is a fact that the various interveners on the market have different convictions. *But if a trader uses such convictions to manage a derived product, he takes the risk that it was possible to avoid* realising a hedge by following the market.

In particular, one takes a risk in assuming that an asset is a martingale. So the logic of the markets is based on the argument of non-arbitrage and the principle of hedging places the question of knowing if the price of an asset is or is not a martingale in the area of *subjectivity*. Consequently, certain works try to know if the markets are efficient theoretically and place themselves according to a super-probability that looks at the world from the point of view of Sirius. This will be a true economic knowledge, but one that nobody has the means to acquire.

Moreover, this problem is not limited to the question of efficiency and certainly concerns a large part of economic theory. If one takes account of the historic and current roles of the economy as an academic discipline on one hand and of the financial markets on the other, one is in the situation where the universal concept suggested by economic science – and in particular the comprehension of the world market – is disqualified by the reality of the market. In other words, we notice in connection with efficiency that certain economic concepts base their objectivity on a very idealised science. Consequently there appears to be a positioning of the function of the economist somewhere between that of scientist on the model of the physicist and that of the engineer or consultant. These use special and precise language in a socio-professional context to construct models or arguments. The pertinence of their representation is indissociable from this premise[25]. We will return much later to the consequence of this development.

If the financial markets challenge the science – in the sense of objective and

---

[25] It is interesting to remark that in the thinking of macro-economics, that despite methodological precautions always conserves a normative aspect, the 1970s, that have so marked finance, have been characterised by the fact that "modelling leaves the academic domain to become an industrial activity with the construction of large models by private institutions (DRI or Wharton) or public ones (BEA)". (P.A. Muet, "Le positif et le normatif dans la modélisation macroéconomique", OFCE, October 1996.

universal knowledge – of every explanation of what the price would be in a few months or of what it must be if there were no speculation, it is because they maintain a particular relationship with knowledge which it is advisable to explain. This brings us to the investigations of new forms of scientific practice.

12

The Special Role of Finance in the Production of Knowledge

The reader will surely be relieved that we have restricted ourselves to a qualitative discourse in order to understand the stakes of contemporary finance, and to be able to tackle its involvement with economy and politics. This needed us to make references to topics from advanced mathematics, such as Brownian motion, Itô calculus, integration, martingales, and so on. This was essential to see clearly and to have a better understanding of the phenomena of risk, the questions of stability and instability, the various forms of speculation, and to understand the epistemological rupture of exact hedging on which the new practices rest.

Clearly this was not difficult. But now critical thought has made its evil way in the scientific domain. The media has difficulty in understanding these areas. There is a resistance that tends to separate scientific expression and the common language into two distinct cultures. One sociologist recently claimed to be able to account for the links between science and society without making any effort to penetrate the contents of knowledge and showed himself in quite a weak light. In the case that we are looking at, this resistance is interesting in itself – it is an index. If it were easy to understand the financial markets, it is probable that their power would not be so great. They would be soon ignored.

Moreover we have only touched lightly on the techniques of finance. We have not entered into the details of the vast growth area of new derived products[1]

---

[1] See "Options (classification)" in the Glossary, page 138.

and their management, nor into the details of methods of taking a position or arbitrage. In these fields there is rapid innovation as much for the products as in the ways of thinking and the models one sees operating in the markets. Ideas renew themselves rapidly[2]. There exist some thirty good international journals on finance, publishing essentially very mathematical articles. But this is only the collective part of the body of knowledge, that only represents a fraction of the available knowledge. For economic, psychological and mathematical speculation the market rooms and their workshops use a body of knowledge that they only divulge when it has no further practical use.

New Scientific Outlets

During the past twenty-five years in the United States, and later in Europe, this new finance has created some scientific outlets. While previously jobs in the banks were principally the privilege of people trained in business studies, economics, or political science, or even literature (e.g. George Pompidou), and only employed engineers for their expertise on the technical health of businesses, there has emerged during the last fifteen years an important outlet for the scientific Grandes Ecoles[3] in France or in other countries, where currently about 15% of students now orient themselves in the direction of finance. These changes in education have also caused changes within the banking organisations. However, new ideas progress quickly.

Brownian motion and Itô calculus are now taught in the schools of finance, while the third year at university provides options in finance linked to the DEA[4] in economic science or probability. This attracts many students. In the economics departments of American universities, at the University of Texas at Austin for instance, information services are made available to students. These are linked continuously to market data (the *spot* prices) on which they can practice the management of a portfolio according to the procedures of hedging being taught. It might be about an exercise title in white sugar or real-time transactions following some protocol. These facilities have been installed, funded by contributions from businesses that also finance the teaching. Such facilities have already made their appearance in France in some Grande Ecoles. The Americans have been quick to realise the importance of teaching these

---

[2] Among the themes of current research we cite: the question of processes presenting jumps, incomplete markets, the evaluation and hedging of complex options, the treatment of rare events and large deviations, etc.

[3] Prestigious French Universities. (Translator)

[4] DEA is a university degree at Masters level. (Translator)

subjects and still have the leading role as creators and exporters of ideas[5].

Teaching and Finance

These new scientific outlets pose a problem for teaching. In every country, with variations, the training of engineers rests essentially on the learning of a *savoir faire* much broader than one generally refers to as the *state of the art*. This consists, above all, of a method of approach to projects that takes account of scientific developments and gives the means for making reasonable decisions. It is a fact that, more and more, the responsibility for decision making is taken by many interveners during the course of a collective laborious childbirth, that requires special training and attenuates the role of the individual originator that is traditionally attached to the trade of engineer. But this does not fundamentally change things concerning the content of the lessons. Finance, on the other hand, knocks over this framework of pre-professional training, that takes its distant origins in the Philosophy of Knowledge, since it is badly suited to providing the knowledge needed by financiers.

The processes of transmission of experience and improvement of ideas in finance are very different from those in industry because the experts are not of the same nature. One cannot imagine diplomas in speculation and its uses. The management of people is more significant in finance than in industry and the clauses in the contracts concerning professional secrets are very strict. The students at schools and universities are not so convinced since they are looking both ways at the same time. This causes the politics of research and development to be *wild*, in the sense that there is a tendency not to reward intellectual work but only its results. Whereas courses on manufacturing processes exist in most engineering schools, a course on the methods of arbitrage is absolutely unthinkable, simply because it would not have any content: if an arbitrage was widely known, the differential on which it was based would disappear and the arbitrage with it. Such courses cannot be conceived outside private organisations connected with financial recruitment establishments, who, as we have seen, have started to improve the creation of experts with more and more expertise.

---

[5] An analogous phenomenon to this occurs in biology within genetic engineering laboratories.

Privatisation of Learning and Knowledge in the Public Domain

Surveys confirm that students and the trainee engineers question more and more the usefulness of their theoretical courses. They find the lessons useless. However, when they consider that what is most useful for them consists of what they know and what others (with whom they are concurrent in the labour market) do not know, one arrives at a situation that tends to divide teaching into two large categories:

– First, the theory, typically a traditionally university topic, accessible to all, free or cheap, founded on documents which are published or at least available. It is about knowledge in the *public domain*[6], where quality is linked to international scientific references in journals, to conferences and to the research undertaken in the scientific communities of each discipline.

– Second, the practical knowledge that is costly and dispensed by practitioners or specialist advisors, who clearly see things from a pre-professional perspective, and who expect payment to share their knowledge.

In the end, one foresees a progressive movement towards a complete privatisation of the body of knowledge, where universal knowledge is synonymous with *without value* but where this knowledge is accessible and valid for all. Alternatively, the body of knowledge has a community, a social class, a corporation, a certain degree of pertinence precisely because of this; according to an inverse hierarchy of values where useful research is kept secret while science in the sense of enlightenment only gains what is left behind.

The same situation does not arise with medical knowledge, for example, where, fearing a progressive privatisation of knowledge, the university research hospital is still pre-eminent. Finance is unique in using knowledge furtively. If a speculator raises an arbitrage that he has discovered, and which then disappears, this is quite different from the situation where a doctor discovers a new treatment. The management of information is crucial in finance, it affects profit. Finance exposes this problem. For all that, the question is more general[7]. Many intellectuals are preoccupied with understanding why culture, not only

---

[6] There is free software called *public domain* that one can obtain and use without a charge. In each category, (word-processing, office tools, scientific calculations, etc.) it represents the lowest rung on the quality ladder. This does not stop their number increasing and their quality improving, but this always lags behind the available commercial software.

[7] I am thinking in particular of the work of M. Gibbons, C. Limoges, H. Nowotny, S. Schwartzman, P. Scott and M. Trow, *The new production of knowledge, the dynamics of science and research in contemporary societies*, SAGE, London, 1995.

scientific but also literary and artistic, is developing at the start of the 21st century without recapturing universality among its reference criteria.

13
Power and Innocence in Finance

The quantity of universal merchandise, and that of a particular merchandise, can be reported in numbers; but the desire to buy and sell is not susceptible to any calculation, and nevertheless the variations of price depends on a moral quantity, which itself depends on opinion and passions[1].

Condorcet

An evolution has occurred over the past fifteen years concerning the role and the power of the state. In France, the governments of the fourth republic, despite their precariousness because of electoral rules that make it difficult to produce majorities, have economic power comparable to that of a monarch in an ancient regime. The creation of money and devaluation are two tools that are at their disposal in a Keynesian context justifying the intervention by the state.

Today, the nation state wields the same power as other decision makers. Through the twin movements of decentralisation and the construction of Europe, the political economy has developed at community, departmental, regional and European level through consultations between local communities and businesses. Public objectives are reached not by the implementation of a legitimate regional power at the ballot box, but more often result from negotiation between the concerned participants – industrial, elected representatives,

---

[1] Condorcet, Lettre au Comte Pierre Verri, 1773, in Oeuvres Complètes, A. Condorcet – O'Connor et F. Arago (Eds), Paris 1847–1849, tome 1, pp285–288.

107

user groups, experts – and between other interested parties[2].

It is no longer sufficient for the realisation of a project, for example a viaduct or tunnel, to undertake an economic study to justify interest in the project. Everyone is aware of the economic relevance of projects, but one prefers to leave the economic agents to play their part in the financial initiative if the project interests them rather than to place their trust in a study made by the services of the state.

Independent of recent developments in finance, it would be sufficient, in the after-effects of the loss of power by the state, taking account of the ambient liberalism, to require in the last resort the structuring of the markets as a new power.

For all that, if the financial markets have power, then it must be analysed.

Are there recent modifications of financial institutions that will explain the development of the past two decades? Can one understand how the decisions that provoke market movements are taken? The question of the psychology of the operators must be addressed first.

Modification of the Influence of Power due to the Appearance of Derivatives Markets

To understand the current role of financial markets in the construction of Europe or the development of the countries of the Third World, it is essential to see how the *creation of the markets for derived products has modified the effect of economic power*. Even the principal players in the economic game, namely the large businesses and the states, are obliged to take account of the signals or the advice of the markets. How has this revolution happened?

On several occasions we have discussed the markets for derivatives, that were introduced into Europe and Japan from the United States in the 1980s and have now spread to emergent markets in other countries by mutation. They constitute a vast growth area that enriches itself each year with several new configurations. Remember that they deal with *futures contracts*. They are concerned with the agreement between two parties to realise later a transaction in return for a payment from one to the other[3]. I have often spoken of options, because these constitute the most important example, but there are also the

---

[2] See notably the work already cited of P. Veltz (see footnote on page 80) where the development of the *French model* is studied and where tools of analysis are proposed in order to understand the current economic operation in the territories and in the world.

[3] Some contracts do not give rise to a payment.

futures, the *swaps*, etc. The word *later* here means "in three months", "in six months" or even, more subtly "whenever you wants before a certain date" or even, with recent products "as soon as such a event happens and before the occurrence of another" etc.

Clearly, many unpredictable events can occur between the agreement and its maturity. It is precisely this that makes these new instruments useful. We have seen for example that an option can allow an organisation (importer or exporter) to get rid of the risks in exchange rates in return for an agreed payment for this service.

To set up the *organised markets* for derivatives – LIFFE in London, MATIF and MONEP in France, DTB in Germany – following the example of the United States was a flash of inspiration. For France, this marked out the international role of Paris. It is important to understand that derivatives correspond to real needs. Proposed by the banks as a service to businesses to let them hedge certain risks, these products quickly became effective tools. But it was not strictly necessary to organise the markets since their underlyings (currency, assets, raw materials, obligations) were already quoted on their organised markets. As we have seen, this allows us to evaluate them, in applying notably the argument of non-arbitrage and to hedge them by following the market.

In creating the derivatives markets – ignoring the usual transactions over the counter – international finance was taking the economic power. How come? By the simple fact that the relationship with the traditional stock market, that essentially only provides an instantaneous quotation of the price, puts into effect a plan that appreciates the much more precise way in which the economy carries numerical judgements over the development of assets and their correlations to the various maturity dates. It is perfectly possible for the markets to convey that a certain currency or asset is going to go up over three months, fall the three months after and then climb again after that. The derivatives markets qualify economic activity in a precise way so that the experts in the Ministry of Finance or the treasuries of large organisations can find here their information on almost every choice they can envisage and *that they cannot ignore* without taking additional risks.

It is clear that if a managing director or government, wants to make their strategy efficient, they must first decide and then try to make good their decision. The methods of expression are thus of great economic importance. In this way, the *derivatives market behaves sociologically like the media*, it expresses public opinion. But if everyone were free to express themselves, it is clearly evident in proportion to his financial pie[4] that each would be able to make

---

[4] The setting-up of derivative markets, and the understanding of the economy is certainly an element to take into consideration if one wants to explain the predominance of the United States at the end of the last century. Firstly, the United

himself understood.

Certain politicians try hard to present this development as tending to constitute absolute power. This seems to me to be excessive, since the economic expertise attached to the markets is *incomplete*, even if it is now much less so than in the past[5]. In particular, the profound factors that will affect the future – quality of education, of scientific research, of social links – are not completely understood. In a general way, and we will return to this, the relative importance of factors affecting the long term is underestimated by the financial markets.

Psychology and the Operation of Markets

From the point of view of an economic operation, a serious drawback to the law of the market is that it has bizarre behaviour. It is a curious phenomenon. One could believe that, for currencies for example, the opening up of the markets together with globalisation would considerably increase the number of interveners and the volume of transactions would lead to a sort of averaging out by a scaling effect. The price would no longer vary much, each intervener being small on the global scale. Physics furnishes us with this intuition, because since the size of a grain of matter is macroscopically significant, its movement must be more regular, there is less thermal agitation and the laws of classical mechanics apply.

The reason is that price has no inertia. The production of goods and the large part of descriptive quantities concerning the volume of the economy displays inertia, since many of the variables concern social phenomena, such as in town planning or in demography. For example, the birth rate of a country like France varies little. The stock market has a soul, which is restless and can panic. To each intervener the market appears to be a meaningful entity – to such a point that the financial press very naturally personalise it: "the market believes that . . . ", "the market is optimistic" and so on – despite this being only the addition of the behaviour of the interveners themselves. At the same time the psychological aspects here are permanent. Through the game of derivatives that watches attentively the anticipations of the participants and by the standard treatment of information (aids to management and the management of

---

States maintains an important conceptual advance over the rest of the world on the techniques linked to derivatives that allows it to innovate. Secondly, the complexities of its techniques allows it to profit by its advances in informatics. Finally, the economic procedures that it was already using, and with which it was familiar, were exported to the entire world.

[5] The mathematicians of finance are perfectly aware of the many research works concerning better modelling of incomplete markets.

portfolio suggesting these recipes are often illusory[6]), their reactions amplify themselves spontaneously and can leave a reasonable situation – speculative bulls – to come back abruptly to a flat calm state as with hysteria after its attack. It is clear that these whims penalise the weak in favour of the strong, for businesses as much as for nations. The president of IMF recently denounced this chronic malady of the system and advocated the creation of special funds to help the member countries who apply sound policies and are not detracted from their route by a temporary loss of confidence of international investors.

Is this an absurd behaviour of the markets? This is a legitimate question.

As psychological aspects are inherent to the operation of the markets, it is suitable to describe them. How do we explain the idea of psychological behaviour of practitioners?

The links between money and emotional life both conscious and subconscious are a subject that clearly needs a great deal of consideration. We will confine ourselves to certain aspects without pretending to be exhaustive. Since the early works of Freud, who described the subconscious interpretation of money in terms of faeces, the recent development of psychoanalysis shows that the effect of money is more complex, not only concerning its role in the form of the cure, but as an identification vector in the organisation of the self[7]. This much larger perspective had however already been opened up by Freud. He wrote that while money was, above all, a means of life and of acquiring power, important sexual factors also played a role in the appreciation of money. Therefore, civilised people might treat questions of money and sex in the same way, that is, with the same duplicity, the same prudency and the same hypocrisy.

The neurotic weaknesses of paranoia and hysteria that are present in more or less marked forms in the psychic system of every individual, can also appear, in a way that helps to orient and mobilise the available energy, like ferments or the catalysts of their activities.

Paranoia is a psychological configuration that tends to make one see the world by interpreting anxieties with global and general ideas where their danger appears to be more dramatic. As for hysteria, it is a delocalised affection, it accords a special place to the particular relation with others. Freud presents it as a factor of subconscious communication. Its mechanism, he says in *The Interpretation of Dreams*, lies in the identification. By this means, the patients express with their symptoms their interior states. They can suffer in various ways and play out alone all the roles of a drama.

---

[6] As we have seen with regards stability in the presence of noise, risk has its illusions as does optics. One believes that such an approach is good because it works nine times out of ten, without thinking that when it runs aground, the loss is nine times greater.

[7] See I. Reiss-Schimmel, *La psychanalyse et l'argent*, Odile Jacob, 1993.

For Freud, the route followed by hysterical imitation is a process of subconscious deductions.

> *Supposing a physician is treating a woman patient, who is subject to a particular kind of spasm, in a hospital ward among a number of patients. He will show no surprise if he finds one morning that this particular kind of hysterical attack has found imitators[8].*

> *... As a rule, patients know more about one another than the doctor does about any of them.*

> *... Let us imagine that this patient had her attack on a particular day; then the others will quickly discover that it was caused by a letter from home, the revival of an unhappy love-affair, or some such thing.*

> *Their sympathy is aroused and they draw the following inference, though it fails to penetrate the consciousness: If a cause like this can produce an attack like this, I may have the same kind of attack since I have the same grounds for having it[9].*

The process of imitation remains subconscious. The deduction does not lead to the worry of encountering the same crisis but to the realisation of the symptom itself.

How do we keep the enthusiasm of the market rooms free from hysteria? There is evidence that there is a lot of subconscious behaviour among financial participants and their propensity to take on the qualms of others is similar to the very psychic mechanism as that described by Freud.

But whereas this search for the qualms of others makes itself apparent through anxiety and qui-vive, the practitioners of financial markets try to interpret the slightest signs of paranoic tones to project an economic significance, that are also present in this behaviour. This thesis was developed by Frédéric Lordon[10] who puts the accent on the interpreted frenzy. Claude Olievenstein[11] showed that paranoia is very much present in daily life and very gradually, just like the jealous husband who reads into the trivial gestures of his wife the proof for which he searches and that confirms his fears, makes one nervous and removed from reality. However, the operators are only human and it is true that the financiers, in their efforts for political power, interpret the prices on the

---

[8] S. Freud, *The interpretation of dreams*, translated by James Strachey, Penguin, 1976, p. 232

[9] S. Freud, op. cit., p. 233.

[10] F. Lordon, "Marchés financiers, crédibilité et souveraineté", revue de l'OFCE, July 1994.

[11] *L'homme parano*, éditions Odile Jacob, 1992.

markets by judging the economy across a grid, that Lordon referred to under the term *credibility*. This grid can be a means for the abuse of power.

We note that the manifestation of both hysteria and paranoia has a self-realising effect but at different levels. On one hand, a paranoiac finds in his reading of the world a confirmation of this, and on the other hand, as Freud emphasised, hysteria is contagious, as are the mental processes at the base in the mobility of capital.

One finds then in the psychological basis for the behaviour of markets some paranoical tendencies and some hysterical tendencies. Is this to say that financiers are more mad than others? Certainly not, finance is simply a very human practice. One will find as much folly and even more in such activities as plastic arts where the work of the avant-garde often proposes an "I exist" incomprehensible as a cry, or in the behaviour of scientists *confounded*.

The question is to know if among the behaviour of the operators of markets, one can describe objectively certain schemes which for to be a spontaneous and natural tendency, are not less dangerous, such that there becomes of them to prescribe some limits.

The analogy with automobile traffic is here very suggestive. Psychology, one knows, plays a large role here: the desire for power, urgency of egotistical needs, and so on. Speed has appeared more and more as a key factor among the behaviour of car drivers. Spontaneous and natural urges generate danger. This has lead to the introduction of a speed limit on the motorway. In the financial markets and more generally in finance, the problem of risk is analogous. As was emphasised by Henry Kaufman at a conference of the International Organisation of Securities Commission,

> The participants in the financial markets are always trying to push back the limits of risks, at least those that they are specifically prevented from taking and that are closely watched[12].

Some national, European[13] and international controls are being progressively put into place. These consist, on one hand, in limiting the lever effect[14] as much as possible by the *prudent ratios* of various kinds, notably in requiring sufficient funds to deal with debts; on the other hand, to instill into the establishments the clear and permanent management of risks by strict internal procedures in order to avoid accidents of the same kind as those at Barings,

---

[12] H. Kaufman, "Ten reasons to reform", Euromoney, November 1992, reprinted in Problèmes Économiques, no. 2347, 1993.

[13] European directive of 1993 on the proper foundation of the establishing of credit.

[14] The derivatives make important lever effects easy. See notably the works of J.A. Scheinkman, of M.H. Miller and of J. Saint-Geours at the colloquium "Risques et enjeux des marchés dérivés" [54].

"accident" which a representative of the supervisory authorities summarised in these terms:

> *How long does it take a bank to discover a rogue trader?*[15]

But this is a formulation to exonerate the leaders. Is Nick Leeson more crazy than you or me? The feverish atmosphere in the market rooms and the pleasurable stimulation that it arouses seems much the same as that experienced at political rallies, on racecourses or in competitive sports: a very human behaviour!

This discussion can clarify the use of the *auction* that we have mentioned in the previous Part. Science imposes more and more rationality in the world of finance through mathematics, economy and by the analysis of risks. Information provided by databases and the communications networks and the methods of calculus also favour this rationalisation; the bank rooms and the organised markets have taken on the appearance of the cybernetic society that science fiction has shown us. But we have known since Fritz Lang and Aldous Huxley about torment in a totally rational world that is transformed into hell. One of the most hysterical expressions of this is the use of the noisy auction at London on the LIFFE following the example of certain American markets. From the bank rooms and large international firms, the networks converge towards the terminals of markets in an amphitheatre like the arteries around its vital heart or else riot in a small noisy and colourful crowd, with its profanity unforeseen as with the Pythian dance at Delphi. What a curious duality! On one hand, science, that is successfully paranoid according to the work of Lacan, and on the other a theatrical improvisation demanding an immediate response that forgoes the reasoning for impulsive intuitions . . . .

It is right to add that no one would wish for a system in which the price of assets would reach successive levels weekly or daily. This would be compatible with economic life, but would prevent the hedging of derivatives. It is thus desirable to favour a certain madness that instantly generates risk.

Moreover, information technology is on its way to conquer the derivatives markets in France, Germany and Switzerland, because it pretends to furnish so much more information to the practitioners.

The Market as Universalism

To appreciate best the place of the financial markets vis-a-vis the rest of the economy, we are going to stay for a while in the field of psychology and show

---

[15] Les Echos of 28/2/95.

that the financial markets are the only ones to operate truly as markets, that
is to say under the single criterion of price level. It will then appear that the
concept of the market is the expression of an ideal.

We are not naturally economic agents. That the exchanges, the work and
the production of goods organise themselves as if each individual was defining
his criteria as a function that rationalises his behaviour, contributing then to
the determination of prices that regulate his transactions, is a very artificial
conception. In the traditional societies whether there is an exchange or not, and
between whom, has often more importance than the result of the transaction
itself[16].

Pierre Bourdieu shows how, for a society like that of the Kabyles[17], the economy as we conceive is opposed to traditional methods. It is above all confined
to women, the men being taken on honour and at good faith, that excludes,
for example, lending, instead of giving, money to one of the family:

> When it is about business, the laws of the family are suspended. Whether
> you are my cousin or not, I treat you as I would any buyer, there is no
> preference, no privilege, no exception, no exemption . . . . It is the opposite
> of the symbolic economy, one calls a spade a spade, and a profit a profit[18].

This situation is analogous, mutatis mutandis, to that of our industrialised
countries. Today, as soon as one looks at something other than the goods of
current consummation, to more complex transactions – property deals, regis-
tered offices and so on, the symbolic economy is entirely in force. For a large
business concerned with the aerial transportation of freight needing the instal-
lation of warehouses and distribution centres across Europe, the countries, the
jobs, the tax system will not be treated in its negotiations as in the markets
but as in areas of meaningful strengths where prestige, vision and the alliances
are significant.

For a long time the economic concept of the market has been criticised for
not taking account of reality. As in the market for work, for example, the classical thesis according to which the level of salaries results from an adjustment
of supply and demand of jobs seems contestable. One can oppose the argument
that the supply of jobs by the managers of businesses is not determined by the
salaries, but by the level of production that they estimate can be sold, from
which the possibility of under-employment at the macro-economic level arises.

---

[16] See on this subject the interesting comparison between a village in India and a
village in Denmark given by Vito Tanzi in "La corruption, les administrations et
les marchés", Finance et Développement, December 1995.

[17] A tribal people, mainly agricultural, based in North Africa. (Translator)

[18] P. Bourdieu, "L'économie des biens symboliques", in *Raisons pratiques, sur la
théorie de l'action*, Seuil, 1994.

Even the markets for accommodation or for transport are obviously governed
by mechanisms much more complex than an equilibrium of the system of prices.

These criticisms have increased more recently with objections of a psychological nature. The health services will never be a market like the others, given
the special relationship between patient and doctor. More generally, the values
that have a tendency to compartmentalise society, as one can see in the United
States, in favouring relations between individuals according to their cultural,
religions or ethnic groups mean that the supply and demand depend on many
categories, whether or not the transaction results in criteria that escape objective analysis. These phenomena seem to confirm one of the theses of the
psychological school of Palo Alto according to which the hypothesis implicit
in every economy that each individual possesses a proper psychic state is rejected and that the only reality is a relational state between one individual and
others[19].

Some come to the same conclusion, like Jean-Pierre Dupuy, by saying that
the economy is a false knowledge:

> *The first characteristic of life in society is influence or mutual dependence,
> the contagion of desires, of feelings and of passions. It is evident for every
> observer of economic life – save perhaps the economists – that the reality
> which we call economic is dominated by these phenomena of influence
> and contagion.*

> *But what one calls economic science forbids us from the start from taking
> account of the mutual influences since it poses some non-socialised beings
> who have independent interests who only establish communication with
> the system of prices[20].*

One often talks a little too readily of the *market* and *economy of the market*.
It is not uncommon to hear talk of a *good* market, but it is a little less common
to see operating a competition that is free and faithful, pure and perfect. The

---

[19] K.J. Arrow, Nobel Prize Winner of 1972, remarked "The point of departure of the
individual paradigm is the simple fact that social interactions come down to interactions between individuals .... A market is, for an economist, the perfect illustra-
tion of a situation resulting from the interactions between individuals." K.J. Arrow "Methodological individualism and social knowledge", American Economic
Review, May 1994. On the Palo Alto school see J.J. Wittezaele and T. Garcia, *A
la recherche de l'École de Palo Alto*, Seuil, 1992; P. Watzlawick and J. Weakland,
Sur l'interaction, Seuil, 1981 and Y. Winkin, *La nouvelle communication*, Seuil,
1981.

[20] J.-P. Dupuy, *Le sacrifice et l'envie, le libéralisme aux prises avec la justice sociale*,
Calmann-Lévy, 1992, see also "Les bases de la théorie économique sont fausses",
exposéau débat "A quoi sert la science économique" ENPC, 1994, Lettre de l'AFSE
July 1994.

report of activities of the Conseil de la Concurrence[21] shows that the balance between the concern of equity and the concern *not to halt the economy* is very difficult to grasp. What is the boundary between cooperation and understanding? The law grows more and more complex. The ethics of concurrence has been tackled in the Charié Report (Assemblée Nationale, December 1993); it proposes principles and appeals to moral and civic values, that appear truly angelical.

These difficulties, by way of contrast, show the financial markets in a new light. They are neutral and free. They do not take account of any particularity of the interveners. They are indifferent to affiliations. They do not care about the fact that the inventions of Citroën are definitive in the history of the motor car, nor that Crédit Lyonnais grew out of the textile industry. From this aspect the financial markets embody the ideal of a neutral equity vis-a-vis cultural relations and attachments.

As Machiavelli showed at the start of the 16th century, the exercise of power is a succession of risks, so that an organisation, in order to perpetuate itself, must possess a plan of action playing the role of suspension such that the train does not overturn with the first few bumps. Just as the divine right of the recognised sovereign preserves the legitimacy of power exercised in fact by its minister, so the democratic ballot exonerates the elector. He is responsible for the consequences of his decisions but knows not to be culpable. And if the financial markets have such power that one fears their tyranny, then there is good reason to believe that their legitimacy rests on an innocence that it is worth understanding.

Firstly, as we have said, free access and universality of prices are two of the most important characteristics of financial markets. They are rightly or wrongly considered as qualities that the other markets do not possess.

Secondly, most of the large interveners on these capitalist markets are not individuals, but pension funds of insurance companies and mutual investment funds. We have already mentioned their cumulative financial importance. It is much greater than that of the treasuries of public funds. These institutional investors gather together the savings of households that are managed by specialists on the market, salaried professionals who have no other objective than to obtain the best returns. It is about a management that deals with the performance of very popular mass savings.

In the end, we could imagine a society where the citizens dedicate a significant part of their incomes and then express their preferences, choices and economic perspectives to responsible managers, by some placements and transactions on diverse markets. The proportion of the remunerations of capital and that of work in the value added to businesses are 35% and 65% respectively in

---

[21] "The Notes Bleues de Bercy", no. 76, December 1995.

countries of the OECD. This ratio varies a little from country to country[22]. So it is theoretically possible for each person to actively manage an account that brings in around a third of his income. Each in his own way will take command of his economy and the markets will surely find an incontestable democratic legitimacy.

It makes good sense however, taking account of chance and risks inherent in the possession of a portfolio of transferable securities, that a sharing of such proportions so egalitarian will not remain stable. On the contrary, in every case one observes a concentration of capital and one runs into the problem of the distribution of wealth. We rack our brains more and more concerning international inequalities. The poorest countries sink, and in the rich countries the gap between those who are inclined towards at least a little sharing[23] and those who are in a precarious situation or are excluded seems to be growing larger and larger.

The power of the financial markets then rests on a certain innocence *provided one forgets the problem of the redistribution of wealth*. That only confers, from a historical view, a semi-legitimate[24] situation, uncomfortable as it is.

However, it seems particularly curious that European construction has given such importance to the financial logic. This is why we devote the rest of this chapter to the European adventure.

European Construction and the Financial Markets

We have seen that the derivatives markets, which appeared in Europe towards the end of the 1980s, have modified the effects of power where they concern the political economy. I would like here to advance the theory that this fact has played an important historical role in the turn of events that have lead to the European construction starting in the 1990s.

During the fifty years that have passed since the premier European institutions created by Jean Monnet and Robert Schuman were put in place, the European construction is characterised by the active role of middle management and by the reticent or disinterested attitude of the general public. So that

---

[22] Cf. Th. Piketty, "L'Économie des inégalités", La Découverte, 1997.

[23] Asked at the start of 1996, 46% of French households estimated they had kept money in 1995 for financial investments (quoted by INSEE, January 1996).

[24] For different points of view on this subject, see the books of S. de Brunhoff, *L'heure du marché*, PUF, 1986 and J.P. Fitoussi and P. Rosanvallon, *Le nouvel âge des inégalités*, Seuil, 1996.

from the start the problem can be stated in the following terms: *how do we translate into fact a general will that does not exist?* This absurd situation[25] can nevertheless change during the modifications of the conditions of life due to the multiplication of economic exchanges that will change our outlook.

It will start with economic measures. The Treaty of Rome creating the Common Market was signed in 1957, the European monetary system was created in 1979. Since then institutions were set up in Brussels and Strasbourg, the standard industries regularised themselves, the movement of products and people was unrestricted and equivalences between professional qualifications were established. Serious difficulties appeared precisely in the area of money. From the creation of the European currency snake, the deutschmark did not stop increasing against the franc which went through six devaluations. It became progressively clear that the convergence of currencies would not be a simple game of margins of fluctuations of the European currency snake, except to harmonise the political economies that were themselves strongly linked to systems of obligatory withdrawals, to regimes of foresight and to right to work, i.e. to social issues.

It was at the start of the 1990s that we ran straight into the problem of political will. How does our outlook develop? Were people brought together? Certainly. But the war lost by the Third Reich has left profound traces. One part of the Danish, the British and the French – made up of intellectuals, as the debate on the ratification of the Treaty of the Union has shown – hardly supports the power of the German economy. But unemployment took grip, generating uncertainty and favouring a general movement with a resurgence of values, local regional and community, at the same time opposed to the political will of European construction[26]

Under these conditions, financial logic appeared on the side of the decision makers as the only recourse to save Europe. The financial logic benefited from the relative legitimacy of the markets which we have talked about. It rests on the incontestable fact that at a time of fluctuation of currencies (and uncertain choice) the creation of a single currency modifies the optional choices of economic participants. More than half the international exchanges are actually made in dollars. As we have explained with the problem of the monetary base, the fact that the businesses of Europe, the first commercial power in the world, use the same unit of accounting, the euro, contributes to fortifying their solidarity that can then redirect economic world order. This lead to the preparation of the Treaty of the Union. To adopt the euro, the candidate countries must

---

[25] Note that the problem of climatic change due to the greenhouse effect, that is caused by the emission of gases into the atmosphere, is posed today exactly in these terms.

[26] To which we add the incapacity, profoundly significant, to organise a strong union in the European ranks.

respect five criteria that concern the fluctuation of money, inflation, budgetary deficit, public debt and interest rates.

Without a doubt the development of financial institutions and the setting-up of derivatives markets created, a little after 1986, a new situation. It became impossible to lead the European construction while the governments remained free to conduct their own political economics. It would even to help to maintain the dynamic of this new sharing out of power to offset the lack of a strong popular European will. This strategy appeared in the report of the Delors Committee (1989), that projected a *European system of central banks* and the convergence of political economies in order to reach a fixed parity. The Maastricht summit in December 1991 pronounced in favour of a single currency. The Treaty of the Union that came into force on 1st November 1993 proposed that central banks should be independent and precludes the creation of a Central European Bank.

*

* *

Today in 2003, at the time of the translation of this book into English, the euro is established as the currency in a dozen countries. The preceding analysis has been largely confirmed. The UK stays outside this single currency club and its interest in the matter is the object of lively political and economic debates.

After an unexpected fall in its first two years, the new currency rests with a group of countries representing an economic weight and has taken a value close to that of the dollar, evolving with a relatively low volatility. The political problems of Europe now retake the foreground, notably the question of a European constitution and an executive power and also that of its enlargement by other candidate countries.

14
Long Term Risks

The World Financial Storm

Agriculture in the countries of the Third World suffers chronically from the *hazards of weather*. One year, drought penalises the production of sugar in Thailand, another year coffee in the Ivory Coast, the following year cotton in Mexico, and so on. To fight against the fates is exhausting, in the end it is against the feeling of helplessness itself that one must fight. To this end, one can perfect irrigation, establish management of water more appropriate to the exigencies of vegetation following the growing phases, fight against parasites, locusts, etc. so that in more normal years it is possible to maintain an output that can be carried to the factories, to be shipped to ports in the buying countries. It is necessary to fix the loading price. The raw materials are quoted, and their transactions make them a function of the market prices.

On the markets it is a different world, where a new hazard reigns that turns upside down the predicted returns to the producing countries. For example, white sugar quoted at \$430 a tonne at the end of 1994 dropped to \$35 in one week in January 1995, a result of information and the movement of capital. This new blow, if one analyses it, has a component linked to the hazards of agricultural production but with a negative correlation: the good years in quantity becomes catastrophes for the price: after a good crop of maize in 1994-95 its price dropped by 20%. Clearly the price is also subject to urges, purely stock market based, that are the repercussion of other markets or the anticipations of traders.

121

The double impact of these hazards compromises the economies of these countries. The total debt of the developing world passed 1810 billion dollars in 1993, and 1930 billion dollars in 1994, while the average public effort from countries of the Committee for Development Aid in favour of these countries dropped from 0.31% to 0.30% during this time.

The situation of small European businesses is similar: they suffer from a financial logic that imposes on them fluctuations of prices and exchange rates. This is costly and favours the present. In the futures, in the game of derivatives, quotations appear day by day. The future is thus projected on to the present where it behaves, without any signs of inertia.

In mechanics, inertia is what makes effects last. The conduct of the economy by the financial markets is like that of a huge ship in a raging sea that the helmsman pilots by remote control from a small boat: the small waves have as much importance as the large swells.

The question of the *development* of financial powers raises itself. During the time of the opposition of two blocks, the idea of an economy without constraint where market forces are freely expressed was a utopian porter of hope opposed to the Marxist utopia. The present period is different and favours a more concrete and detailed way of thinking. In their present form, the markets make poor decisions and ignore the problems of organisation of space and time. A tree takes 30 years to grow, a research laboratory takes 10 years to become effective . . . .

Engineering Culture and Finance Culture

There is a culture that learns about the risks and records them permanently: that of engineering. It has very different characteristics from those of finance. First of all, it endeavours to think what would happen in the event of a breakdown, a rupture, a disaster and how to remedy it. Reliability is essential: the choice of straight thinking that it asks for, the hierarchies of securities that it needs, are a priori of dual concern for the management of technical equipment and industrial production and make the principle grid of analysis for the conception of products. In addition, the technical decisions are not always taken like the ukases[1] of those who have *the knowledge*. The engineers have learned that today decisions are collective acts constructed by dialogue and discussion. These decision processes, from the development of an idea for a project through to its realisation are a work of difficult parturition.

The markets, on the contrary, bring some help in a large measure according

---

[1] A decision that cannot be discussed. (Translator)

to mathematical considerations or absurd psychology, and above all inadvertently destroy brutally without fear of consequences, sometimes without reason.

Let us look at what one does when one designs a bridge. One studies what one calls the *serviceable limit state* and the *ultimate limit state*. The engineer must do this; this double approach is in the regulations:

1. The bridge is designed so that it is not damaged when subject to traffic. There will be maintenance to do, but over the projected duration, the movements and the vibrations that occur must produce no deterioration of its mechanical properties.

2. The engineer also studies what would happen in the event of damage to the bridge. It is clear that there will always be – with small probability – events more serious than one can predict. For reasons of cost and aesthetics, one can only cover the major risks and there will remain some risks not covered. In these extreme conditions, the work will be damaged, and is going to deteriorate more or less: in what way? It is this that the engineer must study.

The limit state of service corresponds to what the work can endure, and one takes account of these bounds for its design. The ultimate limit state corresponds to the complete study of types of damage which could lead to its ruin.

The Eiffel tower was constructed of iron. If it had been made of cast iron, as was more common at that time, it is certain that this would have lead to it being dismantled. Cast iron is a hard material, solid but brittle, so extremely dangerous. Iron has plasticity, and so has steel. If the ultimate limit state of a work in steel is studied carefully, then when an unexpected overloading occurs that creates a constraint greater than it can support in equilibrium, the work will progressively lose its shape and in *continuing to resist*, it will not regain its original form. Such a work twists without breaking. This happens with bridges today, and is what is used to design earthquake-resistant constructions.

The markets, on the contrary, are like sheets of fragile material. They give their warnings too late. They create decisions without taking responsibility for the damage, the limit state.

The development of financial techniques over the past twenty-five years, especially since the 1980s in France, with the sophistication of conceptual tools about which we have spoken, has happened precisely in a period where the role of the engineer in society has been profoundly transformed. In the areas of construction, of traffic or of regional planning, the private interests, the public interests, cultural and symbolic interests, cannot be taken independently of each other. The time has passed when the engineer possessed a regal power given by the public command of the state, even if some like to maintain the nostalgia.

The engineer enters into a dialogue with the various concerned parties. No longer can he cloak himself with the dignity of his discipline and his speciality. More and more he listens, he translates from one language to another, he envisages different variations, he describes the constraints and objectives.

Can we think that these changes are caused by the more liberal and less rigid economy with which we live? There is, without doubt, an acceleration factor but it seems that, more profoundly, one is becoming aware of repeated problems of the environment that have modified the processes of collective decision making, since these questions always present characteristics that make dramatic solutions impossible.

Firstly, they are composed of various categories of interveners according to their implacable ways of thinking (dynamic economics and the creation of jobs, electoral influences and the spreading of populations, etc.).

Next, the situation and its uncertainty about to the *reason to believe* and not the calculation of probabilities, since it is perceived as unique in its history.

Finally the decision will have consequences on its legitimacy and its opportunity (traffic induced by a new road justifies it, etc.).

These characteristics of *complex* situations are taught today to engineers in order to construct a variety of models of the same situation from the perspective of *multiple expertise*[2]. The pedagogy evolved slowly, but clearly. The understanding of complex situations requires a new wisdom that is emerging.

Already the regulations take account of the consequences of decisions, and one cannot today present a project without a study of its impact that takes account as much as possible of events that one might previously have omitted.

The French system with a multiplicity of schools of engineering at the expense of small establishments has hardly adapted to the international scale, but it has the advantage of a large flexibility to adapt its training outlets. Alternative training has become the rule here. In more of the teaching according to classically pedagogy, pupils are made aware of the complexity of the social aspects, by field courses and by case studies where they meet the well-founded basis of contradictory positions of various participants and by projects where they have to acquire conceptual tools and the technical language of applied science needed for discussions. To properly defend a project is to know how to adapt it, modify it and to make it the project for those who will take it over. Young people understand these things very well, as they know that the creation of a dynamic economy that invigorates a region is an objective that if difficult merits flexibility, dialogue and concessions, and that this approach will have a major role in the future.

In this regard, *technical financial thought appears archaic in its relations*

---

[2] See B. Barraqué, "Une expertise différente pour les politiques en réseau", Annales des Ponts & Chaussées, no. 81, 1997.

with society. It develops a specialised knowledge base in a closed system, taking direct hold over the economic power. *The traders stay in front of their computer screens* and like burglars they plan their crimes.

One might think that the problems of the environment are so profoundly new in the history of humanity, that it is likely that there will appear a cultural transition towards a spreading of powers and the limitation of the irresponsibility of the financial market.

Conclusion

The way we have approached finance following the mathematical path gives a new perspective, and allows a more open discourse. The usual approaches take short cuts and follow the ruts where the press, for instance, is only too keen to help. This has cost you a certain technical effort, but I hope I have made you aware of the ideas that history has legitimised: 70 years after Bachelier threw down the challenge, these ideas have been reinvented to meet the needs of market rooms.

We have seen how mathematics has been applied to explain the games of money and chance, with the concept of mathematical martingale, of Brownian motion, and of the integral to represent the speculator's profit, leading then the epistemological rupture of the 1970s. This approach, that does not quite reduce the markets to what certain economists would like them to be, puts strong evidence for the complex links between science and speculation.

Normally when talking about finance, the usual link, namely economic finance, is hampered from a priori points of view and explicit or implicit hypotheses, that mask essential aspects of reality. Speculation is very often confused in the text books with the idea of a forward position. On the contrary, I think that a more realistic view accepts that during the last quarter of the last century, speculation, with its economic, psychological and mathematical dimensions, has become a definite activity, consuming quantities of information and the treatment of information, with clearly noticeable consequences on the resources of businesses and on the jobs held by the senior executives, and their training.

But the financial markets are modifying themselves profoundly: new exchanged products are more varied and the volume of transactions has greatly increased. The discovery of the *exact hedge* by following the market, that we have described in this book, and that makes the markets play the role of insurance companies without constraint over risks, has allowed the world development of derivatives markets. Sprawling and complex, the financial markets furnish instantaneously an enormous number of indicators on the future values and various maturities of quoted assets. In this new context, the transactions must take the greater part of the account of the market itself in the acceptance of risks and their management. To *take account* does not mean to behave absolutely according to the indications of the various instruments of the market, but to adjust as a function of them. A new thinking of the market establishes itself: the various points of view of agents translate into various laws of probability on the development of prices; a principle of relativity links these points of view: it is impossible to know what is good. There is no point of view more legitimate than others, the idea of an objective law of probability is really an abstraction of economic theory. The thinking of the financial markets does not recognise the objectivity of the concepts of *fundamental* or *efficiency* and considers that they are picked up from a political economy and of the corresponding observations of the facts of society.

It is too much to say that the only objective economic reality is the market. Many players study the operation of goods and exchanges, confront their interpretations and their beliefs on what is known, and then construct their fundamentals. It is useful and even indispensable as much at the business level as at the regional and national level. Some institutes exist for this. There are actions which partly make the prices, even if practically one cannot separate them from the speculations we have talked about.

*
* *

This denies a pertinence to every *objective* approach that will have different prospective pretentions of indications of the markets, and this has been the consequence, as we have explained, of the entry of derivatives into the organised markets.

It is interesting to establish a parallel with a philosophy that has historic links with the United States and a renewed interest at the present time. I want to talk about pragmatism.

This movement, founded by the philosopher and logician C.S. Peirce (1839-1914) who gave it the name, was popularised by W. James (1842-1910) and J. Dewey (1859-1952). After a certain decline in the period after the war corresponding to the *glorious thirties*, when it was often the logical positivism that occupied the forefront of American science, it found a significant expansion with H. Putnam and R. Rorty after the 1970s[1]. According to Peirce, to

---

[1] Cf. K. Chatzis, "De Peirce à Rorty: un siècle de pragmatisme", in "Rationalité ou

think of an object is nothing other than to think of all the practical results produced by it. This resolutely non-contemplative philosophy is interested in the action and the procedures of collective justification more than the representations by the individual consciences. The objective reality is not envisaged as an autonomous instance of beliefs and human thought processes. James and Dewey make explicit an instrumental conception of intelligence and the refusal to distinguish between facts and values, between science and ethics[2]. Of the three language fields: the syntax relative to the form, the semantic relative to the reference, the pragmatism that concerns the relation between the speaker and the listener, for Peirce the third is the most important, so much so that the second lost his interest. Rorty gave his authority to the legacy of Dewey and of Nietzsche to *replace theoretical questions with practical questions*[3]. It denies the existence of a real objective, but defends itself nevertheless from all relativism[4] in giving a particular importance to a process of obtaining an intersubjective agreement. It leads then to a certain promotion of science, not for its contents, but for the quality of methods of organisation of the scientific institution:

> *The pragmatists would like to replace the desire of objectivity – the desire*
> *to be in contact with a reality that is more than the community with which*
> *we identify ourselves – by that of solidarity with this community. They*
> *think that the habit of counting on persuasion rather than force, on the*
> *respect of the opinions of their colleagues, the curiosity and passion for*
> *both known and new ideas, are the only virtues possessed by the men*
> *of science. They do not think that there is an intellectual virtue, namely*
> *rationality, superior to moral qualities. [...] On the other hand there are*
> *a large number of reasons to appreciate the institutions that they have*
> *developed and within which their work is effected, giving models for the*
> *rest of culture.*[5].

What then are the essential characteristics of the operation of the scientific

---

pragmatisme?", Annales des Ponts & Chaussées, no. 75, 1995.

[2] A distinction that some consider as a characteristic of the modern period from where the naming of the post-modernists have given sometimes to the neo-pragmatists.

[3] R. Rorty, *Science et solidarité, la vérité sans le pouvoir*, L'éclat, 1990, p. 10.

[4] *"It is sufficient to assimilate the objectivity to the inter-subjectivity to be imme-diately accused of relativism"*, but *"[the adopted pragmatism] the ethnocentric vision where there is nothing to say on truth or rationality if it is not the picture that can be realised of familiar procedures of justification that a given society – ours – puts to work in a determined domain of research."* R. Rorty, op. cit. p. 51.

[5] R. Rorty, op. cit, and he continues *"Thus the institutions and the practices that gather the diverse scientific communities can be taken as a source of suggestions relative to the manner that the rest of the culture can organise in turn."* p. 55.

institution peculiar to making it the ultimate reference of a philosophy that
wants to be without illusions or at least without naivety?

Although experiments and observations play an important role in the natural sciences, it is not through the intermediary of experimental devices on
which the exchange and selection of ideas is founded. The central mechanism
of the scientific institution is the publication in journals using scientists for their
anonymous expertise or *referees*. A considerable number of articles is submitted
and after eventual modification they are published in a journal which is more
or less prestigious. It is crucial in this system that the expert can pronounce
freely because of *keeping secret* all the consequences, negative or positive, and
of the advice which he gives. In fact, his judgement is a judgement of interest,
hence subjective, and his level of exigency depends on the prestige of the journal concerned. An article which is refused by one will be presented to a lesser
journal. In this way everything is done pragmatically. The periodical *Current
Contents* elaborates a hierarchy of reviews by an iterative procedure: the good
authors are the most cited as one can evaluate by the *Science Citation Index*,
and the good reviews are those which are written by the good authors. Admittedly, this system has its problems and difficulties, but these remain anecdotal
considering the colossal number of articles published[6].

One cannot fail to notice a striking parallel between the three current institutions where legitimacy has a sufficient force to institute an actually recognised
historic power. I am talking about the *financial markets*, *democracy* and *science*. They each present two common characteristics: one is abstract namely
the universality at least of the limit state that it more or less approaches; the
other is concrete namely to possess an *escape mechanism*, in the sense that the
clockmaker uses this term, that allows a decision to be taken without which its
responsibility can reflect badly.

As for science, it is not always easy to separate knowledge from knowing
the subject, this being the permanent preoccupation of scientific activity, and
the principal method has been to submit the conclusions to an independent
expert. Just as Archimedes solicited the advice of experts in Alexandra, so the
journals today require the opinion of *anonymous referees*.

We will not elaborate on the universalism of democracy nor on the methods
of representations and ballots. It has many variations and some very elaborate
systems of graded responsibility fashioned by history[7].

Finally for the financial markets, we have seen that they present the characteristics of pure and perfect markets on which the price is the subject of

---

[6] Which currently increases by a factor of about 2.5 every 10 years.

[7] Although it was a regime that it is not proper to call democratic, it is interesting
to note that at Venice, the premier maritime power of the 14th and 15th centuries,
power was held by the doges, who took decisions during secret meetings.

independent decisions which are normally attached to transactions. This is un-
doubtably a virtue in the eyes of liberals. The escape mechanism resides in the
innocence of every operator, even if he is a speculator, whatever the economic
consequences of the movements that he decides.

At this time one can see no credible substitute for these three legitimacies
that make their mark on the modern world. The true question is to know if one
can, and if so in what way, modify, quieten down, and perfect these institutions
such that one can face the problems of the environment and the future of the
planet, while preserving the accumulated experience of civilisation that they
represent. As for science, it needs to find a way to take account of the fact that
it is often dangerous for the environment. Can it continue to create production
techniques without precautions as it did during the 20th century?

Democracy faces new difficulties. It does not progress in the world as one
might wish. In the rich countries of the world we observe that the current procedures do not give politicians the real means to impose on populations those
actions which are necessary in order that future generations are not confronted
with climatic changes and pollution that today we know to be absolutely vital.

As for finance, its brute strength appears to be the most archaic of the three,
in the methods of blindly regulating the problems by players cut off from the
world. The conduct of the economy by the market mechanisms, that we have
seen is equivalent to a voting system based on wealth, poses a serious problem
to each rich country not resolved by the redistribution of resources, and this
induces at the world level egotistical behaviour, indeed desperate behaviour,
hardly compatible with the management of the environment[8]. The measures
envisaged by the IMF to prevent the repetition of the collapse of a country
following a crisis of confidence of the financial markets, such as happened in
Mexico, are based on a different logic. It is not about economic positions and
idealistic financiers: it looks for cooperation. We hope that there is there the
sign of an awareness.

---

[8] The negotiations concerning the problems of climatic changes currently show the
international scene in a dangerous situation of non-cooperation where the United
States is not apparently prepared to see a reduction in its standard of living and
where developing countries which have strong reserves of fossil energy do not wish
to suffer excess emissions of carbon dioxide from the rich countries that are the
reason for their wealth.

Glossary

American option. An option which can be exercised up to its maturity date.

Arbitrage. A word used in the financial domain for a profitable operation *without risk*. By extension, taking account the complexity of risk analysis, arbitrage is often used in the sense of *profit with little risk*. Financial products that have the same characteristic must be negotiated at the same price, or else case profit without risk can be realised. Arbitrage is the identification and exploitation of such anomalies. You see this term in expressions like *no arbitrage opportunity* or *no possibility of arbitrage* or *arbitrage-free price*, a price preventing profit without risk.

For over 70 years, the traders in the financial markets have pushed its consequences in a way that exploits more the reference to the market itself.

For option-pricing and for option-hedging, a single price for the option is one that prevents any possibility of arbitrage as much for the bank as for its client. This supplies the composition of a portfolio that allows exact hedge. This argument for a good price and the corresponding hedging with no opportunity of arbitrage, is general. This goes way beyond the Black-Scholes model, even if, in certain models (underlying jumps, incomplete markets, etc.), this argument is not enough to produce a unique price, and so exact hedging is not possible.

Bachelier, Louis. Born in Le Havre, he taught mathematics at Dijon, Rennes and Besançon. His thesis *Theorie de la speculation* (Theory of Speculation) in 1900, and his memoir *Theorie Mathematique du Jeu* (Mathematical Theory of Games) in 1900, published in the Annales de l'Ecole Normale, are pioneering works both for financial markets, and for the theory of random processes in continuous time. They were developed in his work *Calcul des* *probabilities* (Calculus of Probabilities) (1912). By his treatment of this, which he called *the radiation of probability*, he came to be considered as the founder of the mathematical theory of Brownian motion, and as the forerunner of Markov processes in continuous time.

Black F. and Scholes M. Authors of an article (Black and Scholes, 1973) on the evaluation of options is the birthplace of the new financial methods. This work clearly sits in a stream of thought linking contributions from many other authors (see the historical comments of Merton, 1990, and of Maliaris and Brack, 1982). The model they introduced is more simple than the theory, and is much used (see "Black-Scholes model"). This discovery won them the Nobel Prize for Economics with R.C. Merton in 1997.

Black–Scholes model (1973). The asset is modelled by a random process S_t which is the solution to the equation

$$dS_t = \sigma S_t dB_t + \mu S_t dt$$

where B_t is a Brownian motion. The parameters of the model are the following constants:

$$\sigma = \text{volatility of the asset } S$$
$$\mu = \text{coefficient of the drift}$$
$$r = \text{interest rate}$$

The coefficient μ does not appear in the calculations. The evaluation of an option, for example a call of exercise value K and maturity date T, is the object of an explicit calculation. One finds the stochastic integral that simulates the option

$$k + \int_0^t H(s)dS_s$$

is here

$$C(S_t, t)$$

where the function $C(x, t)$ is the solution to a second order partial differential equation that has in the case of a call, an explicit solution and one which one knows how to solve numerically if not. The deduction of the evaluation formulas and what they indicate how must be composed the hedging portfolio can make different ways. The most luminous mathematically is that which uses Itô stochastic calculus but one can also obtain them in an elementary way by considering that time is made up of discrete pieces. This way is very similar to that followed by Bachelier to establish the *equations on the radiation of probability*. It comes here to approach Brownian motion by a random market.

Brownian motion. Takes its name from the botanist Robert Brown, who observed the disorganised movement of small particles in liquids, the expression today denotes more often the *random process of Brownian motion* that is a random function and which holds a central place in the theory of probability. It is related to other theories such as the propagation of heat or electrostatics. By modelling a price on the stock market by a Brownian motion, as suggested by L. Bachelier, leads to the suggestion that the instantaneous value of the price follows a Gaussian law. This hypothesis has been modified today, (see "Black-Scholes model"). Starting with Brownian motion, a large number of other random processes can be defined which even satisfy principles of a differential calculus: the *Itô calculus*.

Call. An option to buy. A contract giving its owner the right, but not the obligation, to buy an asset at a fixed price at the end of the exercise time.

If this right cannot be exercised until the maturity date T (usually 3 or 6 months, or even much later), the call is said to be *European*. If it can be exercised any time you want before the date T, it is said to be *American*.

Similarly, for an option to sell (a put), we talk of a European put, or an American put. Also see "Put".

Conditional expectation. In any random situation, you can consider the expectation as a whole, or you can consider your expectation taking account that one event is the result of another. This is called *conditional expectation*. For example, if life expectancy in general is 70 years, you could consider the life expectancy of someone who is already 60, or even who is already 80. This idea has been rigorously defined by Kolmogorov, as the best estimate in the least-squares sense, and appears in the definition of mathematical martingales. (See "Martingale.")

Cox–Ross–Rubeinstien (1979). These authors have developed an elementary approach to the discrete–time Black-Scholes model. The interest in this is not only pedagogical, but also algorithmic. Every simulation and numerical method is based on discrete time.

Derivative. Something whose value depends on the development of one or more asset called underlyings. Equivalently, *derivative security* or *contingent claim*.

Efficiency. The efficacity of a market. It can be studied with reference to the power to prevent the appearance of profit without risk, or the capacity to indicate the good economic price.

European option. An option which cannot be exercised until after its maturity date.

Exercise. The *exercise price* is the threshold from which the holder of a call or put to work the right that gave him the option. The *exercise date* of an option is predetermined for a European option, but is open up to the date fixed for an american option.

Fundamental. Sometimes a price that would be if there were no *noise* of agitation in the market, sometimes a price which would be if the true economy was expressed by the market. In one case or the other, this concept is problematic. The logic of the market allows each economic player who gave him a particular content, which is in fact subjective. This does not mean it is not useful.

Future. Contract to buy or sell an asset at some date in the future for a price fixed at the start. It produces a class of derivatives. Futures are governed by specific conditions (dates, units, margins, etc.).

Gaussian hypothesis. The random vector (or multivariate as one says in statistics) which obey the Laplace–Gauss Law, or normal law, have a variance and covariances (in fact moments of all orders) and conserve this type of law under linear transformations.

The random processes which the marginals are normal laws conserve this property under all possible linear transformations. It results that these hypotheses allow the use of very elaborate methods from statistics, from economics (time series) in treating the signal (Wiener filtering) and in calculating work under random requests.

Itô integral. The Wiener integral (see "Wiener integral") applies to the case where the function $f(t)$ depends on the price of the asset up to time t. We say that f is *non-anticipating*. The Itô integral can be defined for more general processes than Brownian motion, the semi–martingales, and allows for the development of a differential calculus whose particular rules make up *Itô calculus*. In this context one can pose and solve differential equations whose solutions are random processes. In the Black-Scholes model, the path of the asset S_t is modelled by the solution of the equation

$$dS_t = \sigma S_t dB_t + \mu S_t dt.$$

Liquidity. The liquidity of a financial asset in a market. It denotes the possibility of buying or selling a large number of units over a short period, without significantly affecting the price of the asset.

Management of options. When a banker sells an option to a client, the sum that he is going to pay at the maturity date is random since it depends on the (unknown) development of the underlying. Nevertheless, it is possible to make up a hedging portfolio which, suitably managed, allows him to dispose of exactly the sum asked (up to errors in the model and the discretisation of the price). The principle is different from that that governs an insurance company. The banker balances the operation with each client and with each underlying.

For European options, in the Black-Scholes model, if $S(t)$ is the price of the underlying , the theory gives the value $F(t, S(t))$ of the hedging portfolio and its composition. The function $F(t, x)$ is the solution to a partial differential equation, with boundary condition $F(T, x)$, where T is the maturity date of the option. In the case of a call (an option to buy),

$$F(T, x) = \begin{cases} x - K & \text{if } x > K, \\ 0 & \text{otherwise.} \end{cases}$$

where K is the exercise value of the option.

To manage the portfolio, one makes use of various partial derivatives of F:

$$\text{delta} = \frac{\partial F(t, S(t))}{\partial x}$$

$$\text{gamma} = \frac{\partial^2 F(t, S(t))}{\partial x^2}$$

$$\text{theta} = \frac{\partial F(t, S(t))}{\partial t}$$

$$\text{vega} = \frac{\partial F(t, S(t))}{\partial \sigma}$$

where σ is the volatility of the underlying.

One says that an option is *covered in delta neutral* if the hedging portfolio contains a quantity of underlying assets such that the state is indifferent to small variations of the market.

American options can be similarly examined. Their theory raises inequalities of partial derivatives.

If the underlying presents jumps, one needs a class of models where only approximate hedging is possible. An evaluation and a best hedge can nevertheless be calculated.

Martingale. In the language of gamblers, the *secret of winning*. In mathematics, a process $M(t)$ with values in a vector space (the real numbers usually), depending on time, such that the value $M(t)$ is the best estimate of $M(t + h)$, given the information available at time t. It is the representation of a game of even chance. Martingales satisfy some important inequalities. Centred Brownian motion is a martingale.

MATIF. Created in February 1986, under the name *Marché a Terme International de France* exemplifies an organised market where one can buy or sell new financial products, negotiable contracts and derivatives (see `http://www.matif.fr`.)

Moment. The expectation of the n–th power of a random variable X with law of probability P is called the *moment of order n*. The moment of order 1 is the *expectation*. The moment of order 2 is the *variance*. The term *moment* comes from mechanics, a probability over R can be considered as a solid bar, and the moment of inertia with respect to the centre of gravity is the variance .

Monte Carlo method. A method for numerical calculation based on the use of random numbers produced by a computer. This method is used in financial mathematics in the case where you don't have an explicit formula to evaluate the hedging of options.

Moving average. The average of values of the price over the last month (or week etc). The moving average varies more regularly than the price itself. This procedure is used in telecommunications or in the treatment of an image in order to filter the signal. One can show that the moving average of a process of Brownian motion type gives no information on its future development.

Option. A means of giving the right, but not the obligation, to buy (in the case of a call) or sell (in the case of a put) an asset underlying at a date and price fixed in advance, in return for a bonus. (See also "Call" and "Put".)

Options (classification). In the same way that in mathematics one often distinguishes between functions and functionals, which are functions of functions, one can distinguish between

1. *Option functions* which are contingent contracts whose definition makes to intervene a certain function of the value of the underlying to the maturity date. These are types of calls and puts and their combinations (straddle, strangle, butterfly, spread, bear spread, etc.) and also digital options which are defined by a jump function (the Heavyside function).

2. *Functional options* whose definition comes from a functional or trajectory of the underlying from the moment before the maturity date. These are a type of barrier option (call cap, put cap, knockout options,

etc.) the asian options (which are made to intervene on the time average of the underlying), look-back options (which are made to intervene on the maximum and minimum attained by the underlying during the period).

3. *Other contingent products* In *over the counter (OTC)* markets one meets made–to–measure products. Among these are non–standard American options (which one cannot exercise with complete freedom during the period) the forward start options, whose period starts at a future date, options involving the exchange of two assets, options with several assets, and so on.

For options which are more or less exotic, see Hull [51].

Organised markets. A market that deals in standard products. A *compensation room* ages at the good end of operations concluded and carries in return of each transaction. It is submissive to a market authority.

The main organised markets are:

BBF	Bolsa Brasileira de Futuros	http://www.embratel.net.br
BELFOX	Belgian Futures & Options Exchange	http://www.belfox.be
CBOT	Chicago Board Of Trade	http://www.cbot.com
CME	Chigago Mercantile Exchange	http://www.cme.com
DTB	Deutsche Börse Group	http://www.exchange.de
HKFE	Hong Kong Futures Exchange	http://hkfe.com
ISE	Italian Stock Exchange Council	http://www.borsaitalia.it
LIFFE	London International Financial Futures and Options Exchange	http://www.liffe.com
MATIF	Marché A Terme International de France	http://www.matif.fr
MEFF	MEFF Renta Fija	http://www.meff.es
SBF	Société des Bourses Françaises, MONEP	http://www.bourse-de-paris.fr
SFE	Sydney Futures Exchange	http://www.sfe.com.au
SIMEX	Singapore International Monetary Exchange	http://www.simex.com.sg
SWX	Suisse	http://www.bourse.ch
TIFFE	Tokyo International Financial Futures Exchange	http://tiffe.or.jp
TSE	Tokyo Stock Exchange	http://www.tse.or.jp.

Out of the money. If the exercise price of a call or put is close the the value of the asset underlying, or more generally if the active zone of an option holds the value of the underlying asset, one says that one is *in the money*. If on the other hand the active zone is far below or far above the current value of the underlying, one is *out of the money*. Such options are more delicate to manage because they are concerned with rare events where the probability is consequently not well–known.

Processes. The terms *random process* and *stochastic process* are synonymous. They denote a random function. Mathematically this is simply a random function taking values in a function space. It results that the law of a process is a probability measure on a function space. The law of a Brownian motion is called a *Wiener measure*.

Put. Option to sell. (See "Call".)

Quadratic variation. For a process it is the sum of the squares of the increases according to a partition of time more and more fine, For a semi–martingale $S(t)$ it exists and is denoted by $[S,S]_t$. One of the definitions for the volatility of $S(t)$ at time t uses $[S,S]$.

Semi–martingale. Semi–martingales are processes more general than martingales, for which one can develop the theory of stochastic integration (Itô calculus) that serve to model the prices. Like the concept of a martingale, this is relative to certain information that clears progressively from the process itself, or even that which clears from the process and other processes.

Swap. When a bank agrees a loan, there is a risk it will not be reimbursed (a credit risk) and the risk of interest rates which come from the fact that the passives are not payed in the same way as actives. For example, when the loan is at a fixed rate, and the bank borrows at a variable rate. To hedge against the risk of rates, the bank can use a swap of interest rates which is an operation equivalent to the realisation with the same compensation of a loan and a borrowing of the same amount, one at a fixed rate and the other at a variable rate.

Transaction cost. To sell an asset in an organised market and buy it back immediately presents a cost, weak, but one that must be counted. These differences between the buying price and the selling price are what we usually call the transaction costs. We distinguish between the asking price, which is the minimum price the operator or market manager is prepared to sell the contract, and the bid price that he is prepared to buy, which is slightly low.

Underlying. An asset under development that carries an option or futures contract.

Variance. The variance of a random variable X is $E[(X - E(X)^2]$. This is equal to $E(X^2) - E(X)^2$. It is the moment of inertia with respect to the centre of gravity of a rod whose mass is distributed according to the law of X. The square root of the variance is the *standard deviation*. The variance of a Brownian motion $B(t)$ is equal to the quadratic variation (see above) from 0 to t and is equal to t.

Volatility. Proportional agitation of an asset. This concept can be tackled in various ways. From the empirical point of view, over a trajectory it could be estimation of the variance or quadratic variation. From the point of view of the management of the portfolio, for the hedging of a derivative, it can be either a parameter of the model allowing the evaluation obtained in an empirical fashion over its underlying, or a consequence of the price on the market for the derivative on which it is based, the family of model being fixed.

In the latter case we speak of *implied volatility* The volatility is usually expressed as a rate with respect to a unit of time, usually a year. For markets of strong currencies, volatilities from 5% to 10% are usual. Volatilities of 15% to 50% or even 100% are encountered for exotic products. The term is not used for the international mobility of capital.

Wiener integral. Suppose, as Bachelier suggested, that the price of an asset is a Brownian motion $B(T)$. What is the algebraic gain from a portfolio that, by buying and selling, constitutes the quantity $f(t)$ of an asset that varies from 0 to T? It is the limit of the sum of the gains made in each small period:

$$f(0)(B(h) - B(0))$$
$$+ f(h)(B(2h) - B(h))$$
$$+ f(2h)(B(3h) - B(2h))$$
$$+ \ldots$$

The limit of this is an integral. It is called the *Wiener Integral* of f with respect to Brownian motion over the interval $[0, t]$.

It is denoted by

$$\int_0^T f(t)dB_t.$$

Witch. A stochastic illusion. Gamblers often believe that if they can see the angels, but they are really witches. For example, the belief that the moving average of a price is increasing and if the price is higher than the moving average, then the tendency is to increase.

Bibliography

[1] M. Aglietta, D.E. Fair and Ch. de Boissieu, *International monetary and financial integration: The European dimension*, Kluwer, 1988.

[2] M. Aglietta, L. Scialom and Th. Sessin, *Prudential supervision reform in Europe, argument and proposals*, University of Wales, Institute of European Finance, 1998.

[3] M. Aglietta, *Macroéconomie financière, Repères*, La Découverte, 1995.

[4] D. Arnould, *Les marchés de capitaux en France*, Armand Colin, 1995.

[5] K.J. Arrow, *Essays in the theory of risk bearing*, North Holland, 1970.

[6] K.J. Arrow and G. Debreu, *Existence of an equilibrium for a competitive economy*, Econometrica, 22, 1954.

[7] P. Artus, *Anomalies sur les marché financiers*, Economica, 1995.

[8] Louis Bachelier, *Théorie de la spéculation, théorie mathématique du jeu*, Ann. Sc. Ecole Normale Sup. 3rd series, vol. 17, 1900. Reprinted Editions Jacques Gabay, 1995.

[9] Louis Bachelier, *Calcul des probabilités*, Gauthier-Villars, 1912. Reprinted Editions Jacques Gabay, 1992.

[10] A. Bensoussan, *On the theory of option pricing*, Acta Applic. Math. 2, 1984.

[11] P. Bernstein, *Capital ideas*, The Free Press, Macmillan Inc., 1992.

[12] F. Black and M. Scholes, *The pricing of options and corporate liabilities*, J. of Political Economy 3, 1973.

[13] Ch. de Boissieu, *Etats-Nations et marchés*, Le Monde, 4 July 1996.

[14] Ch. de Boissieu, D.E. Fair and J. Alwoth, *Fiscal policy, taxation and the financial system in an increasingly integrated Europe*, Kluwer, 1992.

[15] Ch. de Boissieu and D. Lebègue, *Monnaie unique européenne, système monétaire international: vers quelles ambitions?*, PUF, 1991.

[16] N. Bouleau, *Ethique et finance*, Libération, 23 May 1995.

[17] N. Bouleau, *Marchés financiers et économie, un combat á l'issue incertaine*, Les Echos, 17th March 1998.

[18] N. Bouleau, *Finance et opinion* Esprit, November 1998.

[19] N. Bouleau and D. Lamberton, *Residual risks and hedging strategies in Markovian markets*, Stoch. Proc. and Applic. 33, 1989.

[20] N. Bouleau and B. Walliser, *Fuite en avant*, Le Monde des Débats, July-August 1994.

[21] P. Bourdieu, *Practical reason. On the theory of action*, Polity Press, 1998.

[22] P. Bourgine and B. Walliser, (ed.), *Economics and cognitive science*, Pergamon Press, Oxford, 1992.

[23] H. Bourguinat, *La tyrannie des marchés, essai sur l'economie virtuelle*, Economica, 1995.

[24] H. Bourguinat, *Les excès des marchés*, Le Monde des Débats, June 1994.

[25] S. de Brunhoff, *L'heure du marché, critique du libéralisme*, PUF, 1986.

[26] T.E. Copeland and J.F. Weston, *Financial theory and corporate policy*, Addison-Wesley, 1979.

[27] J.C. Cox and M. Rubinstein, *Options markets*, Prentice Hall, 1985.

[28] C. Crouch and W. Steeck (ed.), *Political economy of modern capitalism. Mapping Convergence and Diversity*, Sage, 1997.

[29] R.-A. Dana and M. JeanBlanc-Picqué, *Marchés financiers en temps continu valorisation et equilibre*, Economica, 1994.

[30] P. Devolder, *Finance stochastique*, Ed. de l'Université de Bruxelles, 1993.

[31] M.V. Dothan, *Prices in financial markets*, Oxford University Press, 1990.

[32] L. Dubins and J.L. Savage, *How to gamble if you must*, McGraw-Hill, 1965.

[33] D. Duffie, *Futures markets*, Prentice Hall, 1989.

[34] D. Duffie, *Security markets: stochastic models*, Academic Press, 1988.

[35] D. Duffie, *Dynamic asset pricing theory*, Princeton University Press, Princeton, 1992.

[36] B. Dumas et B. Allaz, *Les titres financiers: équilibres du marché et méthodes d'évaluation*, PUF, 1995.

[37] J.-P. Dupuy, *Le sacrifice et l'envie, le libéralisme aux prises avec la justice sociale*, Calmann-Lévy, 1992.

[38] I. Ekeland, *Le chaos*, Domino Flammarion, 1995.

[39] N. El Karoui, "Modèles de diffusion et introduction aux marchés financiers", Cours de l'Ecole Polytechnique, 1993.

[40] N. El Karoui, M. JeanBlanc-Picqué and St. Shreve, "Robustness of Black and Scholes formula", Mathematical Finance, 8, 1998.

[41] R.J. Elliott and P.E. Kopp, *Mathematics of financial markets*, Springer-Verlag, New York Inc., 1999.

[42] E. Elton and M. Gruber, *Modern portfolio theory and investment analysis*, John Wiley & Sons, 1995.

[43] J.-P. Fitoussi, *Competitive disinflation. The mark and budgetary politics in Europe*, Oxford University Press, 1993.

[44] H. Föllmer and D. Sondermann, "Hedging of contingent claim under incomplete information", in *Applied stochastic analysis*, Davis and Elliott (Eds), Gordon and Breach, 1990.

[45] R. Gibson, *Option valuation*, Mc Graw Hill, 1991.

[46] Ch. Gouriéroux, *ARCH models and financial applications*, Springer-Verlag, New York Inc., 1997.

[47] R. Guesnerie, *L'économie de marché, Dominos Flammarion*, 1996.

[48] R. Guesnerie and J.Ch. Rochet, *(De)stabilizing speculation on future markets. An alternative view point*, DELTA, 1992.

[49] G.J. Holton, *The scientific imagination*, Harvard University Press, Cambridge, Ma., 1998.

[50] C. Huang and R.H. Litzenberger, *Foundations for financial economics*, North Holland, 1988.

[51] J.C. Hull, *Options, futures and other derivativesecurities*, Prentice Hall, 1993.

[52] J. Ingersoll, *Theory of financial decision making*, Rowman and Littlefield, 1987.

[53] INSEE, *Tableaux de l'économie française 1995-1996*.

[54] B. Jacquillat and J.-M. Lasry (Eds), *Risques et enjeux des marchés dérivés*, PUF, 1995.

[55] J.-P. Kahane, "Le mouvement brownien", Actes du Colloque J. Dieudonné, Nice, 1996.

[56] J.-F. Kahn, *La pensée unique*, Fayard, 1995.

[57] J.M. Keynes, *General theory*, Macmillan & Co, 1935.

[58] S.C. Kolm, *Les choix financiers et monétaires*, Dunod, 1967.

[59] D. Kreps, *A course in microeconomic theory*, Princeton University Press, Princeton, 1990.

[60] D. Lamberton and B. Lapeyre, *Introduction to stochastic calculus applied to finance*, Chapman & Hall, 1995.

[61] F. Lordon, "La rationalité en question", Cahiers Français no. 272, July-Sept 1995.

[62] A.G. Malliaris and W.A. Brock, *Stochastic methods in economics and finance*, North-Holland, 1982.

[63] F. Mayor and A. Forti, *Science et pouvoir UNESCO*, Maisonneuve et Larose, 1996.

[64] A.V. Mel'nikov, *Financial markets*, American Mathematical Society, 1999.

[65] R.C. Merton, *Continuous-time finance*, Basil Blackwell, 1990.

[66] "Dostoïevski à la roulette", translated by R.F. Miller and F. Eckstein, Gallimard, 1926.

[67] F.S. Mishkin, *The economics of money*, Banking and Financial Markets, Addison-Wesley

[68] I. Nelken (ed.), *The handbook of exotic options*, Irwin, 1996.

[69] OCDE, *Risques systémiques dans les marchés de valeurs mobilières*, 1991

[70] *Nouveaux défis pour les banques*, 1992.

[71] A. Orléan, "Les désordres boursiers", La Recherche no. 232, May 1992.

[72] J. Perrin, *Les atomes*, PUF, 1948.

[73] E. E. Peters, *Chaos and Order in the Capitals Markets*, John Wiley & Sons, 1991.

[74] H. Poincaré, *Calcul des Probabilités*, rédaction de A. Quiquet, 2ème édition revue et corrigée par l'auteur, Paris, Gauthier-Villars, 1912.

[75] D. Revuz and M. Yor, *Continuous martingales and Brownian motion*, Springer-Verlag, Berlin and Heidelberg, 1990.

[76] S. Ross, "The arbitrage theory of capital asset pricing", J. of Economic Theory, 13, 1976.

[77] D. Ruelle, *Chance and chaos*, Princeton University Press, 1993.

[78] J. Saint-Geours, "Des milliards sans frontières", Le monde des débats, April 1995.

[79] R.J. Schiller, *Market volatility*, MIT press, 1969.

[80] E. Segre, *From X-rays to quarks: Modern physicists and their discoveries*, W.H. Freeman, 1980.

[81] G. Soros, *Le défi de l'argent*, éditions Plon, 1996. See also G. Soros, *Open society: The crisis of global capitalism reconsidered*, Little Brown, 2000.

[82] C. Stricker, "Arbitrage et lois de martingale", Ann. Inst. H. Poincaré, 26, 1989.

[83] L. Tvede, *The psychology of finance, understanding the behavioural dynamics of*, John Wiley & Sons, 2002.

[84] X. de Villepin, "La marche vers la monnaie unique", Rapport du Sénat no. 228, 1994-95.

[85] P. Veltz, *Mondialisation, villes et territoires, l'economie d'Archipel*, PUF, 1996.

[86] B. Walliser, *Anticipations, equilibres et rationalité economique*, Calmann-Lévy, 1985.

[87] Cl. Zaslavsky, *L'Afrique compte! Nombres, formes et démarches dans la culture africaine*, éd. du Choix, 1995.

Index